Foundations of Engineering Mechanics

Series Editors: V. I. Babitsky, J. Wittenburg

Springer

Berlin
Heidelberg
New York
Hong Kong
London
Milan
Paris
Tokyo

A.G. Kolpakov

Stressed Composite Structures

Homogenized Models for Thin-Walled Nonhomogeneous Structures with Initial Stresses

With 26 Figures

Springer

Series Editors:
Vladimir I. Babitsky
Mechanical and Manufacturing Engineering
Loughborough University
Loughborough LE11 3TU, Leicestershire
United Kingdom

Jens Wittenburg
Institut für Technische Mechanik
Universität Karlsruhe (TH)
Kaiserstraße 12
76128 Karlsruhe
Germany

Author:
Professor Dr. A. G. Kolpakov
324, Bld. 95, 9th November str.
Novosibirsk, 630009
Russia

ISBN 3-540-40790-1 Springer-Verlag Berlin Heidelberg New York

Library of Congress Cataloging-in-Publication-Data

A catalog record for this book is available from the Library of Congress.
Bibliographic information published by Die Deutsche Bibliothek.
Die Deutsche Bibliothek lists this publication in the Deutsche Nationalbibliographie;
detailed bibliographic data is available in the Internet at http://dnb.ddb.de

Springer-Verlag is a part of Springer Science+Business Media
springeronline.com

Printed in Germany

Typesetting: camera-ready by author
Cover design: deblik Berlin
Printed on acid free paper 62/3020/M - 5 4 3 2 1 0

Dedicated to my father,
Georgiy Viktorovich Kolpakov,
civil engineer

Contents

Preface

The mechanics of structures with initial stresses is a traditional part of structural mechanics. It is closely related to the important problem of stability of structures. The basic concepts of elastic stability of structures go back to works by Euler (1759) and Bryan (1889). Later, it was found that the problem of deformation of solids with initial stresses is related to variational principles and nonlinear problems in elasticity; see Trefftz (1933), Marguerre (1938), Prager (1947), Hill (1958), Washuzu (1982). Historical detail up to the 1940s can be found in the book by Timoshenko (1953).

Observing the basic concepts of the traditional mechanics of stressed structures, we agree that these are suitable for uniform structural elements (plates, beams, and so on) made of homogeneous materials, but not for complex structures (such as a network plate or a lattice mast) or structures made of composite materials (such as fiber reinforced or textile materials). Many concepts of the classical theory, such as a cross section or neutral plane (axis), correspond to no mechanical objects if we consider an inhomogeneous structure. As a result, we come to the conclusion that it would be useful to have a theory of thin inhomogeneous structures developed on the basis of 3-D elasticity theory with no simplifying assumptions (with no *a priori* hypothesis).

The first problem is to indicate a suitable method for analyzing inhomogeneous structures. The author considers the asymptotic homogenization method and its modifications a suitable method for this purpose. Justification of the homogenization procedure, i.e. the possibility of examining a homogeneous solid instead of the original inhomogeneous composite solid, was one of the principal results of the mathematical theory of homogenization (see Bensoussan et al., 1978; Oleinik et al., 1990; and Zikov et al., 1994). At present, the asymptotic method is applied to many problems involving highly inhomogeneous composites. The asymptotic method uses 3-D models as the starting point, analyzes these models without any simplified hypothesis, and it can be applied to all types of structural elements: solids, plates, rods (see Caillerie, 1984; Kohn and Vogelius, 1984; Trabucho and Viaño, 1987; and Kolpakov, 1991). On the basis of the homogenization theory, engineering analysis methods for composite materials and structures were developed in works by Annin et al. (1993) and Kalamkarov and Kolpakov (1997).

The second problem is what can be taken as the starting model for a body with initial stresses (also called in this book the stressed body). The author takes the 3-D linearized model of an elastic body with initial stresses (see Washizu, 1982) as the starting model and demonstrates that all traditional models of stressed structural elements (composite solids, plates and membranes, beams and strings) can be derived from that model. All the derived models coincide in form with the corresponding classical models, when the corresponding classical model exists. For uniform structures made of homogeneous material, the models derived with the developed method coincide completely with the corresponding classical models.

The new approach to stressed inhomogeneous structures, presented in the book, leads to some significant changes in the concepts of a stressed body (especially, plates and beams). In the classical theory, we incorporate the initial stresses by taking into account the additional forces arising as a result of deformations of stressed structures considered as 1-D or 2-D objects (see Timoshenko and Woinowsky-Krieger, 1959; Vinson, 1989) or analyzing variational principles of the elasticity theory using the classical kinematics hypothesis (see Washizu, 1982). Both methods do not work when we consider complex structures because they assume special local kinematics, while the local kinematics of a complex structure usually has general form. Note that even modern asymptotic methods essentially use the one- or two-dimensional local kinematics (see Ciarlet, 1990; Trabucho and Viaño, 1996). We will analyze structures and incorporate the initial stresses in the frameworks of 3-D elasticity model with no *a priori* assumptions about the local kinematics. The constitutive equation for a body with initial stresses σ^{*}_{jl} can be written (after some transformations described in Sect. 1.1) in the form

$$\sigma_{ij} = h_{ijkl} u_{k,l}$$

with coefficients h_{ijkl} having some symmetry occurring in elastic constants but *not all* of them. This loss of symmetry (directly related to the initial stresses) generates all classical 2-D and 1-D models of stressed structures (plates, membranes, beams, and strings).

The limit models depend on the order of the initial stresses as compared with the characteristic plate thickness or beam diameter. The dependence of parameters of plates and beams (stiffnesses, forces, and moments) on thickness is usual for classical models. Note that in classical models these dependencies arise at the final step of constructing of models.

It should also be mentioned that for composite structures, the distinction between composite materials and multielement structures is lost. For example, a lattice mast can be treated as a beam made of a specific inhomogeneous material. Application of the asymptotic homogenization method makes it possible to approach a "composite material" and a "composite structure" from a common point of view. In this regard, the author does not distinguish in this study between "composite materials" and "composite structures." The approach presented in this book is equally applied to multielements structures and structures made of composite materials.

Although the author of this book pays much attention to the homogenization technique, the subject of the book is not homogenization itself, but mechanics of stressed composite structures. The author considers the homogenization method an effective and useful tool for investigating of mechanical problems related to composite materials and complex structures. The mathematics presented in the book is directly related to the problems of structural mechanics. Taking into account that the homogenization technique is relatively new and not included in engineering courses, the author presents mathematical computations with details.

Chap. 1 contains introductory information. In Sect. 1.1, we discuss the model of elastic body with initial stresses following Washizu (1982). In Sect. 1.2, we discuss two basic methods of the homogenization theory following Sanchez-Palencia (1980) and Marcellini (1979). In Sect. 1.3, the basic ideas of the homogenization method as applied to solids with initial stresses are presented.

Chap. 2 is devoted to composite structures occupying thin regions (plate-like structures). In Sect. 2.1, the homogenization procedure is applied to a 2-D model of a plate with initial stresses. In Sects. 2.2–2.4, we derive 2-D plate models from 3-D model of elastic body. In Sect. 2.5, we approximate 3-D "energy form" by an "energy form" for 2-D plate and derive condition of stability for plates of complex structure. In Sects. 2.6 and 2.7, 2-D membrane model is derived from 2-D and 3-D problems. In Sect. 2.8, 2-D model derived from 3-D elasticity problem is written for a plate with no initial stresses. The important conclusion of Sect. 2.8 is that the plates and membranes of complex structure can be analysed within framework of 2-D models.

Chap. 3 is devoted to composite structures occupying small diameter regions (beam-like structures). In Sect. 3.1, the homogenization procedure is applied to a 1-D model of beam with initial stresses. In Sects. 3.2–3.4, 1-D beam models are derived from 3-D model of elastic body. In Sect. 3.5, we approximate 3-D "energy form" by an "energy form" for 1-D beam and derive condition of stability for beams of complex structure. In Sects. 3.6 and 3.7, 1-D string model is derived from 1-D and 3-D problems. In Sect. 3.8, a 1-D model derived from 3-D elasticity problem is written for a beam with no initial stresses. It is shown that the beams and strings of complex structure can be analysed within framework of 1-D models.

The theory developed in the book (as well as other homogenization theories) becomes practically applicable if one can compute the homogenized constants of structures with no initial stresses. It is why the author devotes Chap. 4 to the computation of homogenized constants of composite structures. In Sects. 4.1 and 4.2, variational principles and bounds for the homogenized stiffnesses of plate-like and beam-like structures are presented. In Sects. 4.3 and 4.4, finite-dimensional approximations of cellular problem are given for plate-like and beam-like structures.

The author wrote Chaps. 2–4, which contain new theories, with details and hopes they are self-containing. The introductory Chap. 1 was written in a condensed style. Detailed exposition of the concept of an elastic body with initial stresses and the two-scale homogenization method (which are presented in Sects. 1.1 and 1.2) can be found in Washizu, K. (1982) *Variational Methods in the Theory of Elasticity and Plasticity*. Pergamon Press, Oxford; and Bensoussan, A., Lions, J.L. and Papanicolaou, G. (1978) *Asymptotic Analysis for Periodic Structures*. North-Holland, Amsterdam. Additional information about the applied problems in the mechanics of composite can be found in the previous book by the author: Kalamkarov, A.L. and Kolpakov, A.G. (1997) *Analysis, Design and Optimization of Composite Structures*. John Wiley. Chichester, New York.

1 Introduction to the Homogenization Method as Applied to Stressed Composite Materials

This chapter contains the basic facts concerning mathematical models of stressed elastic bodies and the asymptotic method of homogenization as applied to elastic solids of a periodic structure. Before formulating the basic problem concerning the mechanics of stressed composites, let us turn to a short survey of the theoretical works for stressed composite materials.

Obviously, the survey should start with the phenomenological model (see e.g. Jones, 1975; Christensen, 1991). As is well known from engineering practice, materials formed from various components (concrete, wool, reinforced plastics, etc.) and various structural elements (perforated and network plates, lattice beams, etc.) can be regarded as homogeneous in a certain sense, and those technical constants can be measured as characteristics of homogeneous materials. In so doing, the characteristics of the materials are defined for a sample sufficiently large as compared with the typical dimension of the components. This approach is in the framework of the so-called phenomenological modeling of composites. Writing the classical linear problem of elasticity for a stressed body with constants and initial stresses defined on the basis of the phenomenological model, one obtains the problem of the stressed composite body. This approach looks reasonable (and even rigorous) from a common point of view. But it does not take into account the inhomogeneity of a structure, which plays a definitive role in the mechanical behavior of a composite. In Sect. 1.3, it will be shown that the phenomenological approach usually is not correct and can lead to an incorrect result for inhomogeneous bodies with initial stresses.

A widely used approach, which can be conditionally called the method of hypothesis, should be mentioned. Models using this approach are based on the introduction of *a priori* assumptions about the stress–strain state of the components of a composite structure. Such kinds models take into account more fully the inhomogeneity of composites. The successful application of analogous approaches in structural mechanics (for example, the classical hypothesis of Bernoulli–Navier, Saint-Venant, Kirchhoff–Love) is a weighty argument in favor of applying the method of hypothesis and for the goals of the mechanics of composite structures. The combination of mechanical common sense with the possibility of experimental verification of the hypothesis has served as the basis for a number of widely used models and engineering methods of analysis of composite materials and structures. At the same time, the negative side of the method of hypothesis is also well known. In it, the author of the model is expected to reliably guess unknown information, which does not always work. The method of hypothesis is widely applied to the analysis of composite materials and structures (see Broutman and Krock, 1974, Kelly and Rabotnov, 1988).

In some specific cases exact solutions can be obtained for 3-D problems in elasticity theory. The number of such solutions is very small.

In the 1970s and 1980s, the so-called homogenization method was elaborated and it was applied to the analysis of composite materials. The foundations of the homogenization theory were laid in the works by Spagnolo (1968), Sanchez-Palencia (1970), followed by papers by De Giorgi and Spagnolo (1973), Berdichevskii (1975), Duvaut (1976), Babuska (1976), Bensoussan et al. (1978), Bakhvalov (1978), Kozlov (1978), Panasenko (1978), and Marcellini (1979). The basic applied directions of the homogenization method are presented in Oleinik et al. (1990), Annin et al. (1993), Trabucho and Viaño (1996), Kalamkarov and Kolpakov (1997), and Cioranescu and Donato (2000). Applications of the homogenization method provided many important results both theoretical and engineering significance. Mention the theoretical prediction of existence of composites with negative Poisson's ratio (Almgren, 1985; Kolpakov, 1985), which were manufactured later (Lakes, 1987; Sigmund, 1994) and application of the homogenization methods to design of composites possessing the required properties (see, e.g., Bendsøe, 1995)

In connection with composites with initial stresses, the so-called "intermediate" homogenization method should be mentioned; it is carried out as follows: one homogenizes the nonhomogeneous body having no initial stresses and calculates the stresses in it by solving the homogenized problem, and then one compiles an operator that should arise in describing a real homogeneous body having those elastic constants and initial stresses in accordance with classical theory. The "intermediate" homogenization method arises in particular from the phenomenological approach to a nonhomogeneous body. In this case, the experimentally measured elastic constants are the homogenized ones. It will be shown later that "intermediate" homogenization in general leads to an incorrect result. Mathematically, this is due to the fact that the G-limit of a sum is not equal to the sum of G-limits (Marcellini (1973)). From the mechanical viewpoint, it is explained by the occurrence of a general state of local stress and strain when the uniform homogenized stresses are applied in a nonhomogeneous medium.

The homogenization method was applied to the analysis of stressed plates; see e.g. Duvaut (1977), Mignot et al. (1980), and Kolpakov (1981, 1987). It should be noted that 2-D equations were taken as the initial model in these works. Therefore, the formulas obtained for effective characteristics in these works are applicable only to those plates whose thickness is significantly less then the length of the period of the structure. This condition, however, is not always fulfilled. Quite often, the thickness and the period of the structure are comparable in magnitude. Under these conditions, it is evident that 3-D distribution of inhomogeneities should be taken into account in calculating the effective properties. The homogenization problem for periodically inhomogeneous plates, in the case where the thickness of the plate and the length of the period are small and comparable in order, was examined by Caillerie (1984) and Kohn and Vogelius (1984). Plates having no initial stresses were studied in the papers mentioned.

The homogenization problem for a periodically inhomogeneous beam, in the case where the diameter of the beam and the length of the period are small and comparable in order, was examined by Trabucho and Viaño (1987), Tutek and Aganovich (1987), and Kozlova (1989) for a beam of coaxial structure and by

Kolpakov (1991) for a beam of general periodic structure. In the paper mentioned, 3-D elasticity equations were taken as the initial model. The papers mentioned dealt with beams with no initial stresses.

Analysis of inhomogeneous materials having initial stresses on the basis of the homogenization method applied to a 3-D model of an elastic body started with the paper by Kolpakov (1989). In the papers by Kolpakov (1989, 1990, 1992b, 2001), asymptotic analysis was carried out for 3-D composite solids and structures. In the paper by Kolpakov (1995a), the homogenization method was applied to a body of small thickness, and a 2-D model for a stressed inhomogeneous plate was obtained. In the papers by Kolpakov (1992a, 1995b), a stressed elastic body of small diameter was investigated, and a 1-D model for a stressed beam was obtained. In the paper by Kolpakov and Sheremet (1997), 2-D membrane and 1-D string models were derived from 3-D elasticity problem.

1.1 Linear Model of a Stressed Elastic Body

In this section, we present a model of a stressed elastic body which will be used as a basic model in our investigations.

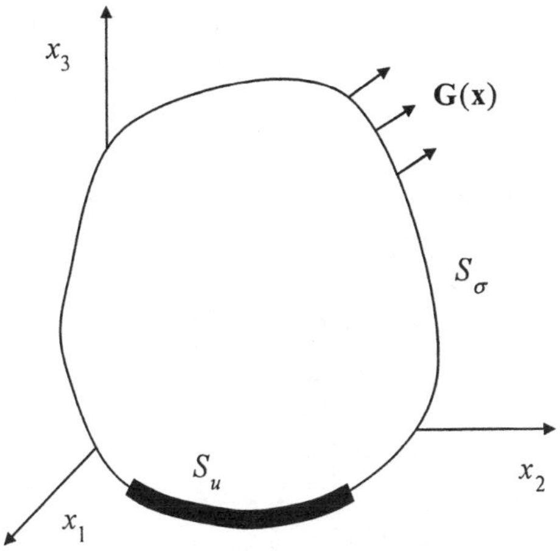

Fig. 1.1. Elastic body

As initial stresses, we understand the stresses σ^*_{ij} existing in a body in its initial state, i.e. before the deformation we are considering. These stresses can be the result of incomparable deformations (for example, dislocations), thermal deformations, or (the case of the most practical interest) forces applied to the body. For the problem with initial stresses, the initial state is taken as the original state.

Consider a body occupying a domain G and clamped along a surface S_u (Fig. 1.1). Let the initial stresses σ^*_{ij} be the result of the application of mass forces $\mathbf{F}(\mathbf{x})$ and surface forces $\mathbf{G}(\mathbf{x})$ on the surface S_σ (Fig. 1.1).

In this case, the initial stresses σ^*_{ij} satisfy the following equations:

$$\sigma^*_{ij,j} + F_i(\mathbf{x}) = 0 \quad \text{in } G, \tag{1.1.1}$$

$$\sigma^*_{ij} n_j = G_i(\mathbf{x}) \quad \text{on } S_\sigma, \tag{1.1.2}$$

$$\mathbf{u} = \mathbf{u}^0(\mathbf{x}) \quad \text{on } S_u. \tag{1.1.3}$$

In this section, the subscript $, j$ means $\partial / \partial x_j$.

Following Washizu (1982), we call the initial stresses σ^*_{ij} satisfying (1.1.1) and (1.1.2) self-equilibrium stresses.

In applying additional mass forces $\mathbf{f}(\mathbf{x})$ and surface forces $\mathbf{g}(\mathbf{x})$ to the body, the problem of the deformation of a body having initial stresses appears. Denote the additional stresses (that is, the stresses according to the additional deformation) by σ_{ij}.

Following Washizu (1982), we apply the principle of virtual work according to stresses $\sigma_{ij} + \sigma^*_{ij}$. This principle can be written as follows:

$$\int_G (\sigma_{ij} + \sigma^*_{ij}) \delta e_{ij} d\mathbf{x} = \int_{S_\sigma} (\mathbf{G}+\mathbf{g}) \delta \mathbf{u} d\mathbf{x} + \int_G (\mathbf{F}+\mathbf{f}) \delta \mathbf{u} d\mathbf{x}, \tag{1.1.4}$$

where strains e_{ij} are computed through displacements $\mathbf{u}$ in accordance with the relations of the geometrically nonlinear theory of elasticity:

$$e_{ij} = \frac{1}{2}(u_{i,j} + u_{j,i} + u_{k,i} u_{k,j}), \tag{1.1.5}$$

and δ indicates the variation of the corresponding functions (see, e.g., Courant and Hilbert, 1953).

Using (1.1.1) and (1.1.2), we can transform (1.1.4) into the following form:

$$\int_G (\sigma_{ij} \delta e_{ij} + \sigma^*_{ij} u_{k,i} \delta u_{k,j}) d\mathbf{x} = \int_{S_\sigma} \mathbf{g} \delta \mathbf{u} d\mathbf{x} + \int_G \mathbf{f} \delta \mathbf{u} d\mathbf{x}. \tag{1.1.6}$$

Under conditions that the displacements **u** are small and the initial stresses $\overset{*}{\sigma}_{ij}$ are not large, we obtain from (1.1.6) the following variational principle for a body having initial stresses:

$$\int_G (\sigma_{ij}\delta e_{ij} + \overset{*}{\sigma}_{ij} u_{k,i} \delta u_{k,j}) d\mathbf{x} = \int_{S_\sigma} \mathbf{g}\delta\mathbf{u} d\mathbf{x} + \int_G \mathbf{f}\delta\mathbf{u} d\mathbf{x}, \tag{1.1.7}$$

where the strains $e_{ij}(\mathbf{u})$ are computed in accordance with the relations of the geometrically linear theory of deformations:

$$e_{ij}(\mathbf{u}) = \frac{1}{2}(u_{i,j} + u_{j,i}), \tag{1.1.8}$$

$$\sigma_{ij} = c_{ijkl} e_{kl} . \tag{1.1.9}$$

Here, c_{ijkl} is a tensor of elastic constants.

The variational principle (1.1.7) leads to the following model of a linear elastic body having initial stresses $\overset{*}{\sigma}_{kj}$: equations of equilibrium,

$$(\sigma_{ij} + \overset{*}{\sigma}_{kj} u_{i,k})_{,j} + f_i(\mathbf{x}) = 0 \quad \text{in } G, \tag{1.1.10}$$

boundary conditions,

$$(\sigma_{ij} + \overset{*}{\sigma}_{kj} u_{i,k}) n_j = g_i(\mathbf{x}) \quad \text{on } S_\sigma, \tag{1.1.11}$$

$$\mathbf{u} = \mathbf{u}^0(\mathbf{x}) \quad \text{on } S_u \tag{1.1.12}$$

with the constitutive equations (1.1.9).

Transformation of the model of a stressed elastic body

Consider the expression

$$\sigma_{ij} + \overset{*}{\sigma}_{kj} u_{i,k}, \tag{1.1.13}$$

which we see in (1.1.10) and (1.1.11). Using (1.1.9), we can write (1.1.13) as

$$\sigma_{ij} + \overset{*}{\sigma}_{lj} u_{i,l} = c_{ijkl} u_{k,l} + \overset{*}{\sigma}_{lj} \delta_{ik} u_{k,l} = (c_{ijkl} + \overset{*}{\sigma}_{lj} \delta_{ik}) u_{k,l} .$$

Here and afterward, $\delta_{ii} = 1$, and $\delta_{ik} = 0$ when $i \neq k$.

Introducing

$$h_{ijkl} = c_{ijkl} + \sigma^*_{jl}\delta_{ik} \,, \tag{1.1.14}$$

we can write the equilibrium equations (1.1.10) and the boundary conditions (1.1.11), (1.1.12) in the following form: equations of equilibrium,

$$(h_{ijkl}u_{k,l})_{,j} + f_i(\mathbf{x}) = 0 \quad \text{in } G, \tag{1.1.15}$$

boundary conditions,

$$h_{ijkl}u_{k,l}n_j = g_i(\mathbf{x}) \quad \text{on } S_\sigma, \tag{1.1.16}$$

$$\mathbf{u} = \mathbf{u}^0(\mathbf{x}) \quad \text{on } S_u \,. \tag{1.1.17}$$

Usually initial stresses σ^*_{ij} are determined from the solution of the linear elasticity problem and they are symmetrical with respect to indexes i and j.

Proposition 1.1. *If the initial stresses* σ^*_{ij} *are symmetrical with respect to* i *and* j, *then the quantity* h_{ijkl} (1.1.14) *has the following symmetry*

$$h_{ijkl} = h_{klij} \,.$$

Generally,

$$h_{ijkl} \neq h_{ijlk} \,.$$

Proof. In accordance with definition (1.1.14), $h_{ijkl} = c_{ijkl} + \sigma^*_{jl}\delta_{ik}$. By virtue of the symmetry of the elastic constants c_{ijkl}, stress tensor σ^*_{jl}, and tensor δ_{ik}, $h_{ijkl} = c_{ijkl} + \sigma^*_{jl}\delta_{ik} = h_{klij}$.

We consider $h_{ijkl} = c_{ijkl} + \sigma^*_{jl}\delta_{ik}$ and $h_{ijlk} = c_{ijlk} + \sigma^*_{jk}\delta_{il}$. Generally, $\sigma^*_{jl}\delta_{ik} \neq \sigma^*_{kj}\delta_{li}$, thus, $h_{ijkl} \neq h_{ijlk}$.

The model (1.1.15)–(1.1.17) will be a basis for the following investigations.

1.2 Homogenization Method in the Mechanics of Composites. The Basic Approaches

As was noted in the preface, the theory presented in this book is based on the application of the homogenization method to a 3-D model of an elastic body with initial stresses. We a give brief introduction to the basic approaches used in the homogenization method.

The approaches to the homogenization problem for composites of a periodic structure are based on different mathematical techniques. The basic directions are presented in Bensoussan et al. (1978), Sanchez–Palencia (1980), Spagnolo (1968), Marcellini (1973,1975), and Jikov et al. (1994). Presented here are two approaches, which will be used in our investigations. They are the asymptotic expansions method and the G-limit approach. It should be noted that these methods are equivalent to one another for many cases. In particular, these methods are equivalent for elastic composites; see Kalamkarov and Kolpakov (1997).

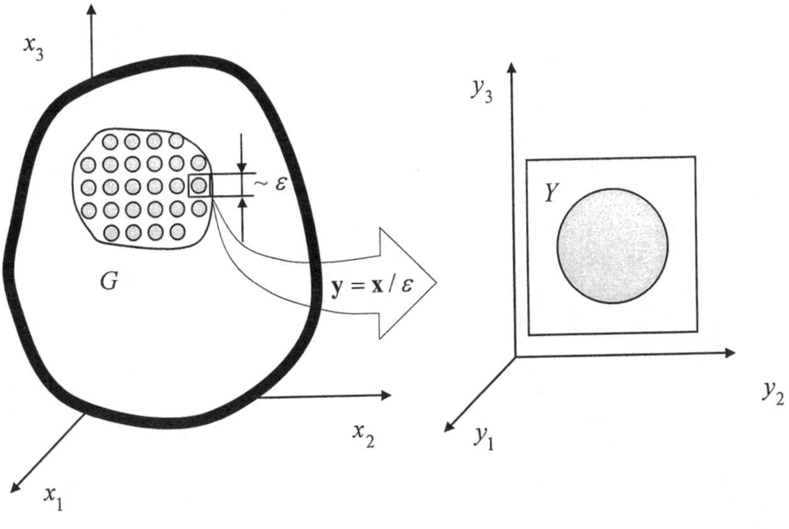

Fig. 1.2. A composite body of periodic structure and its periodicity cell Y in fast variables

Consider an inhomogeneous elastic body with a regular distribution of elastic properties; see Fig. 1.2. The problem of the theory of elasticity for that body has the form

$$L_\varepsilon \mathbf{u}^\varepsilon + \mathbf{F}(\mathbf{x}) = 0 \quad \text{in } G, \tag{1.2.1}$$

$$\mathbf{u}^{\varepsilon}(\mathbf{x}) = 0 \quad \text{on } \partial G , \tag{1.2.2}$$

where

$$L_{\varepsilon}\mathbf{u} = [c_{ijkl}^{\varepsilon}(\mathbf{x})u_{k,l}]_{,j} \; . \tag{1.2.3}$$

Here G designates the region occupied by the composite, ∂G designates its boundary, L_{ε} is the elasticity theory operator, $\mathbf{u}^{\varepsilon}$ are the displacements, and $c_{ijkl}^{\varepsilon}(\mathbf{x})$ is a tensor of elastic constants. The body is clamped on the boundary ∂G. We assume throughout the book that the surface ∂G is sufficiently regular to apply the basic theorems from Lions and Magenes (1972), and Ekeland and Temam (1976), in particular, the formula of integrating by parts.

Here ε is a parameter, which will be associated with the characteristic dimension of inhomogeneity of a composite (see Fig. 1.2). Thus, we consider a problem with parameter (or, in other words, a sequence of elasticity problems).

The following standard conditions are applied to the elastic constants $c_{ijkl}^{\varepsilon}(\mathbf{x})$ (see, e.g., Fichera, 1972): for all $\mathbf{x} \in G$,

$C1.$ $a)$ $c_{ijkl}^{\varepsilon}(\mathbf{x}) = c_{klij}^{\varepsilon}(\mathbf{x})$,

$b)$ $c_{ijkl}^{\varepsilon}(\mathbf{x}) = c_{ijlk}^{\varepsilon}(\mathbf{x})$;

$C2.$ $\left| c_{ijkl}^{\varepsilon}(\mathbf{x}) \right| < M$;

$C3.$ $c_{ijkl}^{\varepsilon}(\mathbf{x})e_{ij}e_{kl} > m\left\| e_{ij} \right\|^2$ for all e_{ij} such that $e_{ij} = e_{ji}$,

where the constants $0 < m,M < \infty$ do not depend on $\mathbf{x}$, e_{ij} and $\varepsilon > 0$,

$$\left\| e_{ij} \right\|^2 = \sum_{i,j=1}^{3} e_{ij}^2 \; .$$

The uniform boundary condition (1.2.2) does not restrict our consideration. It is known from Bensoussan et al. (1978), and Jikov et al. (1994) that the homogenized constants of elastic composites do not depend on the type of boundary conditions. We consider here uniform boundary conditions for simplicity. Boundary conditions of other types will be considered in Sects. 2 and 3.

Let us note that problem (1.2.1)–(1.2.3) permits the following variational formulation (see, e.g., Gajevski et al.,1974 and Washizu, 1982): find $\mathbf{u}^{\varepsilon} \in \{H^1(G)\}^3$ from the solution of the minimization problem,

$$J_{\varepsilon}(\mathbf{u}) - < \mathbf{F}, \mathbf{u} > \rightarrow \min , \tag{1.2.4}$$

where

$$J_{\varepsilon}(\mathbf{u}) = \frac{1}{2} \int_G c^{\varepsilon}_{ijkl}(\mathbf{x}) e_{ij}(\mathbf{u}) e_{kl}(\mathbf{u}) d\mathbf{x} , \tag{1.2.5}$$

$$e_{ij}(\mathbf{u}) = \frac{1}{2}(u_{i,j} + u_{j,i}) .$$

In this book, $\{H^1(G)\}^3$ means closure according to the norm

$$\|\mathbf{u}\|_1 = \sum_{i=1}^{3} \int_G \sum_{j=1}^{3} (|u_{i,j}|^2 + |u_i|^2) d\mathbf{x} \tag{1.2.6}$$

of a set of finite functions, which are infinitely smooth in G:

$$C_0^{\infty} = \{\mathbf{u}(\mathbf{x}) \in \{C^{\infty}(G)\}^3 \ : \ \mathbf{u}(\mathbf{x}) = 0 \text{ for } \mathbf{x} \in \partial G\} .$$

The pointed branches in (1.2.4) signify dual coupling of the elements from $H^{-1}(G)$ and $H^1(G)$ in the standard duality of these spaces. The dual coupling coincides for sufficiently smooth functions with a scalar product in $L_2(G)$. For further details, see Lions and Magenes (1972) and Ekeland and Temam (1976). The equivalence of problems (1.2.1), (1.2.2), and (1.2.4) is a well-known result (see, e.g., Ekeland and Temam, 1976 and Washizu, 1982).

Problem (1.2.1) and (1.2.2) can be written in the following (so-called weak form; see Lions and Magenes, 1972)

$$\int_G \sigma^{\varepsilon}_{ij} v_{i,j} d\mathbf{x} = \int_G \mathbf{F}\mathbf{v} d\mathbf{x} \quad \text{for any } \mathbf{v} \in H^1(G), \tag{1.2.7}$$

where

$$\sigma^{\varepsilon}_{ij} = c^{\varepsilon}_{ijkl}(\mathbf{x}) u^{\varepsilon}_{k,l} , \tag{1.2.8}$$

and $\mathbf{v} \in H^1(G)$ is the trial function.

A homogeneous body (specifically, one which we want to put in correspondence with a composite) is described by the problem

$$Lu = (C_{ijkl}u_{k,l})_{,j} = F_i \quad \text{in } G, \tag{1.2.9}$$

$$\mathbf{u}(\mathbf{x}) = 0 \quad \text{on } \partial G, \tag{1.2.10}$$

or by the minimization problem: find $\mathbf{u} \in \{H^1(G)\}^3$ from the solution of the minimization problem,

$$I(\mathbf{u}) - \langle \mathbf{F}, \mathbf{u} \rangle \to \min, \tag{1.2.11}$$

where

$$I(\mathbf{u}) = \frac{1}{2} \int_G C_{ijkl} e_{ij}(\mathbf{u}) e_{kl}(\mathbf{u}) d\mathbf{x}. \tag{1.2.12}$$

Here, C_{ijkl} are homogenized elastic constants describing a homogeneous material. It is clear that the constants depend on the local elastic constants of the composite.

The coupling of operators L_ε and $L(\sigma)$ and functionals I_ε and I is well known (see, for example, Gajewski et al., 1974). The functionals I_ε and I are the potentials of the corresponding operators, and operators L_ε and $L(\sigma)$ are derivatives, in the sense of Gateaux (Ekeland and Temam, 1976), of the corresponding functionals.

Asymptotic expansion based approach to the analysis of media with a periodic structure

Periodic materials and structures are widely utilized in engineering practice. They are less prevalent in nature, which depends not on the absence of periodicity in natural materials and objects but, as a rule, on the combination of periodicity with various small derivations from it. One meets a lot of periodic structures among plate-like and beam-like engineering structures (lattice covers, masts, frames, and so on).

Let us consider the case when an inhomogeneous elastic body has a periodic structure in coordinates x_1, x_2, x_3, with a period (called periodicity cell, unit-cell, or basic-cell) $\varepsilon Y = [0, \varepsilon T_1] \times [0, \varepsilon T_2] \times [0, \varepsilon T_3]$ (see Fig. 1.2). The factor ε is similar to the characteristic dimension of the periodicity cell (see Fig. 1.2). The material characteristics of the indicated medium are described by periodic functions in spatial variable functions of the following type (see Bensoussan et al., 1978)

$$c^{\varepsilon}_{ijkl}(\mathbf{x}) = c_{ijkl}(\mathbf{x}/\varepsilon), \tag{1.2.13}$$

where $c_{ijkl}(\mathbf{y})$ are periodic functions with a periodicity cell Y. The functions of the argument $\mathbf{x}/\varepsilon$ describe media of the kind considered (periodic with rapidly oscillating characteristics).

The method of asymptotic expansions is based on the ideas of solving the problem, including rapidly oscillating coefficients $c_{ijkl}(\mathbf{y})$ in the form of the following special series:

$$\mathbf{u}^{\varepsilon} = \mathbf{u}^{(0)}(\mathbf{x}) + \varepsilon \mathbf{u}^{(1)}(\mathbf{x},\mathbf{y}) + \ldots = \mathbf{u}^{(0)}(\mathbf{x}) + \sum_{k=1}^{\infty} \varepsilon^{k} \mathbf{u}^{(k)}(\mathbf{x},\mathbf{y}), \tag{1.2.14}$$

where $\mathbf{y} = \mathbf{x}/\varepsilon$ is a "fast" variable and $\mathbf{x}$ is a "slow" variable, i.e. a two-scale expansion is considered. Functions $\mathbf{u}^{(k)}(\mathbf{x},\mathbf{y})$ in (1.2.14) are assumed to be periodic in variable $\mathbf{y}$ with periodicity cell Y. Function $\mathbf{u}^{(0)}(\mathbf{x})$ is a function only of "slow" variable $\mathbf{x}$. By substituting $\mathbf{y}$ for $\mathbf{x}/\varepsilon$, these functions become periodic in $\mathbf{x}$ with periodicity cell εY.

We will seek the solution of problem (1.2.1), (1.2.2) in the form of an asymptotic expansion (1.2.14).

While differentiating, we will separate the variables according to the formula

$$\frac{\partial Z}{\partial x_n} = Z_{,nx} + \varepsilon^{-1} Z_{,ny}. \tag{1.2.15}$$

Here and afterward, superscript $,nx$ means $\partial/\partial x_n$, and the superscript $,ny$ means $\partial/\partial y_n$.

The elasticity theory operator L_{ε} on the left-hand side of equation (1.2.1), allowing for the differentiating rule (1.2.15), can be written as

$$L_{\varepsilon}\mathbf{u} = (\varepsilon^{-2} A_1 + \varepsilon^{-1} A_2 + A_3) \sum_{k=0}^{\infty} \mathbf{u}^{(k)} + \mathbf{f}(\mathbf{x},\mathbf{y}) = 0, \tag{1.2.16}$$

where

$$A_1\mathbf{u} = [c_{ijkl}(\mathbf{y}) u_{k,ly}]_{,jy},$$

$$A_2\mathbf{u} = [c_{ijkl}(\mathbf{y}) u_{k,ly}]_{,jx} + [c_{ijkl}(\mathbf{y}) u_{k,lx}]_{,jy},$$

$$A_3\mathbf{u} = c_{ijkl}(\mathbf{y})[u_{k,lx}]_{,jx} \ .$$

Equating the terms of the same order in ε in (1.2.16), we obtain an infinite sequence of problems; the first three have the following form:

$$A_1\mathbf{u}^{(0)} = 0 \ , \tag{1.2.17}$$

$$A_1\mathbf{u}^{(1)} + A_2\mathbf{u}^{(0)} = 0 \ , \tag{1.2.18}$$

$$A_1\mathbf{u}^{(2)} + A_2\mathbf{u}^{(1)} + A_3\mathbf{u}^{(0)} + \mathbf{f}(\mathbf{x},\mathbf{y}) = 0 \ . \tag{1.2.19}$$

Equation (1.2.17) is satisfied identically because function

$$\mathbf{u}^{(0)} = \mathbf{u}^{(0)}(\mathbf{x}) \tag{1.2.20}$$

does not depend of the variable $\mathbf{y}$; see (1.2.14).

By virtue of (1.2.20), (1.2.18) takes the form

$$[(c_{ijkl}(\mathbf{y})u^{(1)}_{k,ly}]_{,jy} + [c_{ijkl}(\mathbf{y})]_{,jy}u^{(0)}_{k,ly} = 0 \ . \tag{1.2.21}$$

By separating the variables **x** and **y**, the solution of (1.2.21) can be set up as

$$\mathbf{u}^{(1)}(\mathbf{x},\mathbf{y}) = \mathbf{X}^{kl}(\mathbf{y})u_{k,lx}(\mathbf{x}) + \mathbf{V}(\mathbf{x}) \ , \tag{1.2.22}$$

where $\mathbf{X}^{kl}(\mathbf{y})$ represents a solution of the problem

$$\begin{cases} [c_{ijmn}(\mathbf{y})X^{kl}_{m,ny} + c_{ijkl}(\mathbf{y})]_{,jy} = 0 \quad \text{in } Y, \\ \mathbf{X}^{kl}(\mathbf{y}) \text{ is periodic in } \mathbf{y} \text{ with periodicity cell } Y \ . \end{cases} \tag{1.2.23}$$

We will call problem (1.2.23) a cellular problem for a periodicity cell. It also is called a basic-cell or a unit-cell problem. Problem (1.2.23) is the elasticity problem for 3-D composites. Later, we will introduce cellular problems for composite plates and beams.

Let us consider equations (1.2.19), in which the function $\mathbf{u}^{(2)}$ is unknown and $\mathbf{x}$ is a parameter. This problem has a periodic solution if the following equality is fulfilled:

$$\langle A_2 \mathbf{u}^{(1)} + A_3 \mathbf{u}^{(0)} \rangle + \langle \mathbf{f} \rangle(\mathbf{x}) = 0, \tag{1.2.24}$$

where

$$\langle \circ \rangle = (mes\ Y)^{-1} \int_Y \circ\, d\mathbf{y}$$

indicates the average value over periodicity cell Y. The average value $\langle\rangle$ should not be confused with the dual coupling $<,>$.

From the homogenized equation (1.2.24), on account of (1.2.22), we obtain the homogenized (called also average or macroscopic) equation (1.2.9) for $\mathbf{u}^{(0)}(\mathbf{x})$ with the boundary conditions (1.2.10).

For the coefficients C_{ijkl}, the following formula can be obtained from (1.2.22) and (1.2.24):

$$C_{ijkl} = \langle c_{ijmn}(\mathbf{y}) X^{kl}_{m,ny}(\mathbf{y}) + c_{ijkl}(\mathbf{y}) \rangle. \tag{1.2.25}$$

It is known (it can be proved at a mathematical level of rigor, see, e.g., Bensoussan et al., 1978) that

$$\mathbf{u}^{\varepsilon}(\mathbf{x}) \to \mathbf{u}^{(0)}(\mathbf{x}) \text{ , as } \varepsilon \to 0 \text{ weak in } H^1(G), \tag{1.2.26}$$

where $\mathbf{u}^{\varepsilon}(\mathbf{x})$ is the solution of the initial problem (1.2.1), (1.2.2) and $\mathbf{u}^{(0)}(\mathbf{x})$ is the solution of the homogenized problem (1.2.9), (1.2.10). It also known that

$$\mathbf{u}^{\varepsilon}(\mathbf{x}) - [\mathbf{u}^{(0)}(\mathbf{x}) - \varepsilon \mathbf{X}^{kl}(\mathbf{y}) u^{(0)}_{k,lx}(\mathbf{x})] \to 0 \text{, as } \varepsilon \to 0 \text{ in } H^1(G). \tag{1.2.27}$$

From relations (1.2.8) and (1.2.27), the following approximation for a stress field can be obtained

$$\sigma^{\varepsilon}_{ij} \approx [c_{ijmn}(\mathbf{y}) X^{kl}_{m,ny}(\mathbf{y}) + c_{ijkl}(\mathbf{y})] u^{(0)}_{k,lx}(\mathbf{x}). \tag{1.2.28}$$

Formula (1.2.28) can be written in the form,

$$\sigma^{\varepsilon}_{ij} \approx S_{ijkl}(\mathbf{x}/\varepsilon) e^{(0)}_{kl}(\mathbf{x}),$$

where

$$S_{ijkl}(\mathbf{y}) = c_{ijmn}(\mathbf{y}) X^{kl}_{m,ny}(\mathbf{y}) + c_{ijkl}(\mathbf{y}),$$

is the so-called the tensor of local stresses and

$$e_{ij}^{(0)}(\mathbf{u}) = \frac{1}{2}(u_{i,j}^{(0)} + u_{j,i}^{(0)})$$

is the homogenized strains tensor.

It is known (see Bensoussan et al., 1983, Oleinik et al., 1985) that

$$\langle \sigma_{ij}^{\varepsilon} \rangle = \sigma_{ij}, \tag{1.2.29}$$

where $\langle \sigma_{ij}^{\varepsilon} \rangle$ are the average values of local stresses $\sigma_{ij}^{\varepsilon}$ and σ_{ij} are the stresses determined from the solution of the homogenized problem,

$$\sigma_{ij} = C_{ijkl} u_{k,lx}^{(0)} = C_{ijkl} e_{kl}^{(0)}. \tag{1.2.30}$$

The stresses σ_{ij} determined by equality (1.2.30) are called homogenized stresses. One can derive formula (1.2.29) by averaging (1.2.28) over periodicity cell Y with regard to the definition of the homogenized constants (1.2.25).

It follows from (1.2.26) that in the limit (as $\varepsilon \to 0$), a nonhomogeneous material will behave like a homogeneous elastic material with effective coefficients of elasticity C_{ijmn}, given by (1.2.25). But, it is seen from (1.2.28) and (1.2.30), local stresses $\sigma_{ij}^{\varepsilon}$ can differ (and usually they differ) from homogenized stresses σ_{ij}. As seen from (1.2.29), the averaged values $\langle \sigma_{ij}^{\varepsilon} \rangle$ of the local stresses (but not the local stresses themselves) coincide with the homogenized stresses σ_{ij}. Note that the real (substantively existing) stresses in a composite are local stresses $\sigma_{ij}^{\varepsilon}$.

Technical issues concerning the application of the asymptotic expansion method and its modification for 2-D (plate-like) periodic structures and 1-D (beam-like) periodic structures are presented in the subsequent sections of the book.

G-limit based approach to analysis of media with a periodic structure

The G-limit approach is a sophisticated mathematical method used in the mathematical analysis of homogenization problems and proof of convergence theorems. It is not used (at least at present) in applied homogenization, except for a few special cases, one of which will be presented in Sect. 1.3. Nevertheless, it must be mentioned in a book related to homogenization because of its important role in the theory of homogenization. We required certain mathematical notations, which were introduced in the works by Spagnolo (1967) and Marcellini (1973).

Let us denote by V the Banach reflexive and V^* a space topologically conjugated with V (see Ekeland and Temam, 1976). Following Marcellini (1973), let us denote by $C_0(V)$ a set of convex, functionals in V, which assume taking values in $(-\infty,+\infty]$, not identically equal to $+\infty$, and semicontinuous from below. Let us also denote

$$C(\alpha, v_0, M) = \{F \in C_0(V) \ : \ F(v) \le \alpha(v) \text{ for all } v \in V, \ \ F(v_0) \le M < \infty\},$$

where the functional $\alpha \in C_0(V)$ is such that $\alpha(v) - \langle v^*, v \rangle$ reaches a minimum in V for any $v^* \in V^*$.

The functional F^*, defined in V^* through the equality

$$F^*(v^*) = \sup_{v \in V} [\langle v^*, v \rangle - F(v)]$$

is called a conjugate to F.

Definition 1. *The sequence of functionals* $\{F_\varepsilon\} \in C(\alpha, v_0, M)$ *G-converges to the functional F, if for any* $v^* \in V^*$, $\lim_{\varepsilon \to 0} F_\varepsilon^*(v^*) = F^*(v^*)$.

Definition 2. *The sequence of operators* $L_\varepsilon : V \to V^*$ *G-converges to the operator* $L : V \to V^*$, *if operators* $\{L_\varepsilon\}$ *and* L *are invertible and for any* $v^* \in V^*, L_\varepsilon^{-1} v^* \to L^{-1} v^*$ *weakly in V for* $\varepsilon \to 0$.

The equivalence of these two definitions of G-convergence was proved in the work of Boccardo and Marcellini (1976).

Note that the conjugated functionals for elastic solids can be associated with the elastic energy of solids and G-convergence can be interpreted in a terms of the convergence of the elastic energy of an elastic composite with the elastic energy of a homogenized solid. A detailed analysis of the convergence of energy will be given for composite plates and beams in Chaps. 2 and 3.

1.3 Homogenization Method in the Mechanics of Stressed Composites

The following conclusion of the previous section is important for further investigations: the local stresses in inhomogeneous media are rapidly oscillating functions with periodicity cell εY. We denote the local initial stresses by $\sigma_{ij}^{*\varepsilon}$, where ε means the characteristic dimension of oscillation.

The problem of the theory of elasticity for an inhomogeneous linearly elastic body under the influence of initial stresses $\overset{*}{\sigma}{}^{\varepsilon}_{ij}$ can be written in the following form [see (1.1.13) and (1.1.14)]:

$$L_\varepsilon \mathbf{u}^\varepsilon + M_\varepsilon \mathbf{u}^\varepsilon + \mathbf{F} = 0 \quad \text{in } G, \tag{1.3.1}$$

$$\mathbf{u}^\varepsilon(\mathbf{x}) = 0 \quad \text{on } \partial G, \tag{1.3.2}$$

where

$$L_\varepsilon \mathbf{u} = [c^\varepsilon{}_{ijkl}(\mathbf{x}) u_{k,l}]_{,j} \;, \tag{1.3.3}$$

$$M_\varepsilon \mathbf{u} = [\delta_{ik} \overset{*}{\sigma}{}^{\varepsilon}_{jl}(\mathbf{x}) u_{k,l}]_{,j} \;. \tag{1.3.4}$$

We consider the uniform boundary condition (1.3.2) that does not restrict the generality of our investigations (other types of boundary conditions will be considered in Chaps. 2 and 3).

The initial stresses $\overset{*}{\sigma}{}^{\varepsilon}_{ij}$ are to be assumed satisfying equations (1.1.1) and (1.1.2). They can be computed in accordance with (1.2.28). Then, they are rapidly oscillating functions.

Problem (1.3.1) and (1.3.2) permits the following variational formulation: find $\mathbf{u}^\varepsilon \in H^1(G)$ from a solution of the minimization problem,

$$J_\varepsilon(\mathbf{u}) + I_\varepsilon(\mathbf{u}) - <\mathbf{F}, \mathbf{u}> \to \min, \tag{1.3.5}$$

where

$$J_\varepsilon(\mathbf{u}) = \frac{1}{2} \int_G c^\varepsilon_{ijkl}(\mathbf{x}) e_{ij}(\mathbf{u}) e_{kl}(\mathbf{u}) d\mathbf{x} \;, \tag{1.3.6}$$

$$I_\varepsilon(\mathbf{u}) = \frac{1}{2} \int_G \delta_{ik} \overset{*}{\sigma}{}^{\varepsilon}_{jl}(\mathbf{x}) e_{ij}(\mathbf{u}) e_{kl}(\mathbf{u}) d\mathbf{x}, \tag{1.3.7}$$

$$e_{ij}(\mathbf{u}) = \frac{1}{2}(u_{i,j} + u_{j,i}). \tag{1.3.8}$$

Problem (1.3.1) and (1.3.2) can be written in a weak form,

$$\int_G \sigma_{ij}^{\varepsilon} v_{i,j} \mathbf{dx} = \int_G \mathbf{Fvdx} \quad \text{for any } \mathbf{v} \in \{H^1(G)\}^3,$$

where

$$\sigma_{ij}^{\varepsilon} = [c_{ijkl}^{\varepsilon}(\mathbf{x}) + \delta_{ik}\sigma_{jl}^{*\varepsilon}(\mathbf{x})]u_{k,l}(\mathbf{x}). \tag{1.3.9}$$

Equation (1.3.9) can be written in the form,

$$\sigma_{ij}^{\varepsilon} = h_{ijkl}(\mathbf{x})u_{k,l}(\mathbf{x}),$$

where $h_{ijkl} = c_{ijkl}^{\varepsilon}(\mathbf{x}) + \delta_{ik}\sigma_{jl}^{*\varepsilon}(\mathbf{x})$, and considered as a local governing equation relating additional stresses $\sigma_{ij}^{\varepsilon}$ to elastic constants $c_{ijkl}^{\varepsilon}(\mathbf{x})$, initial stresses $\sigma_{jl}^{*\varepsilon}(\mathbf{x})$, and derivatives of displacements $\mathbf{u}$. The principal difference between the elastic constants $c_{ijkl}^{\varepsilon}(\mathbf{x})$ and the coefficients $h_{ijkl}(\mathbf{x})$ is related to that symmetry. Coefficients h_{ijkl} do not have all symmetry occurring in the elastic constants (see Proposition 1.1 and condition $C1$).

A homogeneous material (specifically, one which we want to put in correspondence with a composite) is described by the problem

$$L(\sigma)\mathbf{u} = [C_{ijkl}(\sigma)u_{k,l}]_{,j} = -f_i(\mathbf{x},\mathbf{y}) \quad \text{in } G, \tag{1.3.10}$$

$$\mathbf{u}(\mathbf{x}) = 0 \quad \text{on } \partial G,$$

or by the problem in variational form: find $\mathbf{u} \in \{H^1(G)\}^3$ from a solution of the minimization problem,

$$J_{\sigma}(\mathbf{u}) - <\mathbf{f},\mathbf{u}> \rightarrow \min, \tag{1.3.11}$$

where

$$J_{\sigma}(\mathbf{u}) = \frac{1}{2}\int_G C_{ijkl}(\sigma)e_{ij}(\mathbf{u})e_{kl}(\mathbf{u})\mathbf{dx}. \tag{1.3.12}$$

Here, $C_{ijkl}(\sigma)$ are homogenized constants describing a homogeneous material. In the case under consideration, the constants depend on both the local elastic constants of the composite and the local initial stresses.

The coupling of operators L_{ε}, M_{ε}, and $L(\sigma)$ and functionals I_{ε}, J_{ε}, and J_{σ}, is the following: functionals $I_{\varepsilon} + J_{\varepsilon}$ and J_{σ} are the potentials of the operators

$L_\varepsilon + M_\varepsilon$ and $L(\sigma)$, and operators $L_\varepsilon + M_\varepsilon$ and $L(\sigma)$ are derivatives, in the sense of Gateaux, of the corresponding functionals.

The homogenization theory presented in Sect. 1.2 provides a method for calculating of $C_{ijkl}(0)$ (i.e. constants for a body having no initial stresses). To analyze the problem under consideration, we must modify the homogenization method while taking into account the initial stresses.

Asymptotic expansion based approach to analysis of media of a periodic structure with initial stresses

Let us consider a medium with a periodic structure. As above, we will designate a periodicity cell of the structure by εY (see Fig. 1.2).

The material characteristics of the indicated medium are described by periodic functions in spatial variables of the type

$$c^{\varepsilon}_{ijkl}(\mathbf{x}) = c_{ijkl}(\mathbf{x}, \mathbf{x}/\varepsilon), \tag{1.3.13}$$

$$\overset{*}{\sigma}{}^{\varepsilon}_{kl}(\mathbf{x}) = \overset{*}{\sigma}_{kl}(\mathbf{x}, \mathbf{x}/\varepsilon),$$

where $c_{ijkl}(\mathbf{x}, \mathbf{y})$ and $\overset{*}{\sigma}_{kl}(\mathbf{x}, \mathbf{y})$ are periodic functions of variable $\mathbf{y}$ with a periodicity cell Y. The functions of the argument $\mathbf{x}/\varepsilon$ describe media of the kind considered (periodic with rapidly oscillating characteristics and rapidly oscillating initial stresses).

The method of asymptotic expansions based on the ideas of solving the problem in the form of a special series (1.2.14) can be used to investigate a problem of this kind.

The asymptotic expansion method for structures that occupy thin regions, i.e. inhomogeneous plates and membranes, is described in Chap. 2, and Chap. 3 describes structural members with a small cross-sectional area, i.e. inhomogeneous beams, rods, and strings.

G-limit based approach to analysis of media with a periodic structure and the "intermediate homogenization" procedure

The homogenization method discussed in Sect. 1.2 can be applied to operators with coefficients

$$h_{ijkl}(\mathbf{x}, \mathbf{x}/\varepsilon) = c_{ijkl}(\mathbf{x}, \mathbf{x}/\varepsilon) + \delta_{ik}\overset{*}{\sigma}_{jl}(\mathbf{x}, \mathbf{x}/\varepsilon), \tag{1.3.14}$$

describing a body with initial stresses; see Sect. 1.1, (1.1.13) and (1.1.14).

At first sight, the problem consists simply of calculating the homogenized constants $C_{ijkl}(\sigma)$, which correspond to operators with coefficients of (1.3.14). However, the problem of computing of the homogenized (G-limit) operator is not a trivial problem, and has no solution for now.

The first proposition, how to compute the homogenized characteristics corresponding to the sum (1.3.14), which seems to be obvious from a common point of view, is the following: let us take

$$C_{ijkl}(\sigma) = C_{ijkl}(0) + \delta_{lk}\langle \overset{*}{\sigma}_{jl}(\mathbf{x},\mathbf{y})\rangle , \tag{1.3.15}$$

or, that is the same,

$$L(\sigma)\mathbf{u} = L(0)\mathbf{u} + M\mathbf{u} , \tag{1.3.16}$$

where

$$L(0)\mathbf{u} = [C_{ijkl}(0)u_{k,l}]_{,j} , \tag{1.3.17}$$

$$M\mathbf{u} = (\delta_{ik}\langle \overset{*}{\sigma}_{jl}\rangle u_{k,l})_{,j} . \tag{1.3.18}$$

Here, $C_{ijkl}(0)$ mean the homogenized constants of the body with no initial stresses: $C_{ijkl}(0) = C_{ijkl}$ [coefficients C_{ijkl} were introduced by (1.2.25)].

Problem (1.3.10) is equivalent to the following problem of minimization:

$$J_{\sigma}(\mathbf{u}) = J_0(\mathbf{u}) + I_0(\mathbf{u}) - \langle \mathbf{f}, \mathbf{u}\rangle \to \min , \quad \mathbf{u} \in \{H^1(G)\}^3 , \tag{1.3.19}$$

where

$$J_0(\mathbf{u}) = \frac{1}{2}\int_G C_{ijkl}(0)e_{ij}(\mathbf{u})e_{kl}(\mathbf{u})d\mathbf{x} , \tag{1.3.20}$$

$$I_0(\mathbf{u}) = \frac{1}{2}\int_G \sigma_{jl}e_{ij}(\mathbf{u})e_{kl}(\mathbf{u})d\mathbf{x} . \tag{1.3.21}$$

In accordance with (1.2.29), $\langle \overset{*}{\sigma}_{jl}\rangle = \sigma_{jl}$, where σ_{jl} are the homogenized stresses.

The right-hand side in (1.3.15) arises when we use so-called “intermediate” homogenization, which is carried out as follows: we homogenize the nonhomogeneous body having no initial stresses and calculate the stresses in it by solving problem (1.2.9), (1.2.10), and then we compile an operator that should arise in describing a real homogeneous body having those elastic constants and initial stresses in accordance with the classical theory presented in Sect. 1.1. The “intermediate” homogenization arises in particular from a phenomenological approach to a nonhomogeneous body. In this case, the experimentally measured elastic constants are the homogenized ones. The coincidence of the experimentally measured and the homogenized characteristics for composites with no initial stresses (for stressed composites, this is not correct) was explained in detail in the book by Bakhvalov and Panasenko (1989).

However, in fact the situation is not so simple. As was mentioned, determining the effective properties $C_{ijkl}(\sigma)$ is equivalent to calculating the G-limit of the sequence of operators given by the left-hand side of (1.3.14) for $\varepsilon \to 0$. In doing so, it is necessary to express the G-limit through limits (of any kind) of operators with coefficients $c_{ijkl}(\mathbf{x}/\varepsilon)$ and $\overset{*}{\sigma}_{jl}(\mathbf{x},\mathbf{x}/\varepsilon)$, that is, we are required to calculate the G-limit of a sum of operators, $\mathrm{G}-\lim\limits_{\varepsilon\to 0}(L_\varepsilon + M_\varepsilon)$, where L_ε and M_ε introduced by (1.3.3) and (1.3.4) are differential operators of the same order. Equivalently, we are required to calculate the strong limit for $\varepsilon \to 0$ of the sequence of functionals $\lim\limits_{\varepsilon\to 0}(I_\varepsilon + J_\varepsilon)^*(v^*)$, where I_ε and J_ε are potentials of the operators L_ε and M_ε given by (1.3.6) and (1.3.7), respectively. Little is known of the methods for calculating the G-limits of the sum of operators and functionals in general. However, it is known (see Marcellini, 1973, 1975) that a G-limit conversion generally does not have the property that "the limit of the sum is equal to the sum of the limits." Thus, the form of the connection between elastic constants and initial stresses of type (1.3.14) is lost through an averaging description of solids. This is the reason for the appearance of the difference between the dependence of the elastic properties of inhomogeneous solids and the same dependence for homogeneous solids.

Therefore, "intermediate" homogenization in general leads to an incorrect result. Mathematically, this is due to the fact that the G-limit of a sum is not equal to the sum of G-limits (see Marcellini, 1973). From the mechanical viewpoint, it is explained by the occurrence of a general state of local stress and strain when uniform homogenized stresses are applied in a nonhomogeneous medium. This example demonstrates that the interaction between the inhomogeneity and initial stresses can lead to effects, which have no analogs in the mechanics of homogeneous solids.

2 Stressed Composite Plates and Membranes

This chapter is devoted to inhomogeneous stressed structures occupying thin regions (plate- and membrane-like structures). Theories based on 2-D and 3-D models are presented.

The homogenization method was applied to the analysis of periodically inhomogeneous plates (with no initial stresses) by numerous investigators; see Artola and Duvaut (1977), Duvaut (1977). The homogenization problem for periodically inhomogeneous thin plates having initial stresses was examined by Mignot et al. (1980), and Kolpakov (1981, 1985, 1987). In the papers mentioned, 2-D equations of the theory of thin inhomogeneous plates were taken as the initial ones. Therefore, the homogenized models obtained in these works are applicable only to those plates whose thickness is significantly less then the length of the period. This condition, however, is not always fulfilled. Quite often, the thickness and the period of the plate are comparable in magnitude. Under these conditions, it is evident that 3-D distribution of inhomogeneities should be taken into account.

The homogenization problem for periodically inhomogeneous plates, in the case where the thickness of the plate and the length of the period are compatible in order, was examined by Caillerie (1984) and Kohn and Vogelius (1984). In the papers mentioned, a modification of the asymptotic homogenization method was proposed. It consists of the direct application of the formalism of two-scales expansion to the spatial problems for a thin layer. The definition of "fast" coordinates in the tangential direction was associated with a rapid periodic change in the material characteristics. The transverse coordinate (in which there is no periodicity) was associated with the small thickness of the layer. In these papers, a problem with no initial stresses was considered. The problem of a stressed thin elastic layer was examined by Kolpakov (1995a).

We will consider nonhomogeneous plates of periodic structure (which are widely used in modern structures). The nonhomogeneity of the plate can be a result of both the nonhomogeneity of the material from which the plate is made, and the nonhomogeneity of the plate geometry. Both cases will be covered by our consideration. Classical uniform plates are a partial case of the plates under consideration.

2.1 Stressed Plates (2-D Model)

As was noted for plates whose thickness is significantly less than the length of the period, 2-D equations of the theory of thin inhomogeneous plates may be taken as the initial one. In this section, we briefly outline the asymptotic expansion method as applied to a 2-D model of an inhomogeneous plate with a periodic structure.

Formulation of the problem

We will examine a plate of periodic structure obtained by repeating a certain small periodicity cell εY over the Ox_1x_2 plane (Fig. 2.1). Here ε is the periodicity cell's characteristic dimension, which is assumed to be small (that is, formalized in the form $\varepsilon \to 0$).

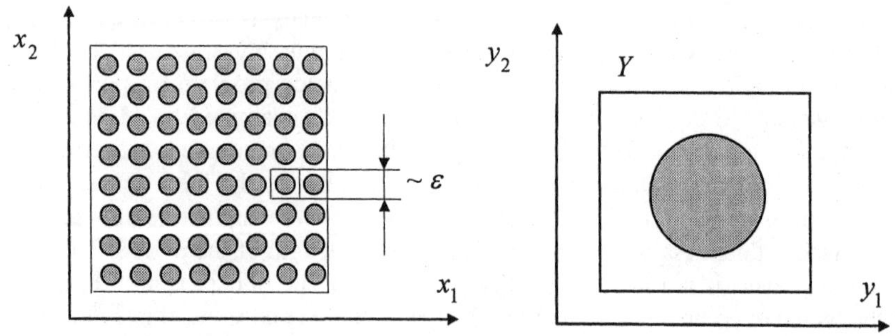

Fig. 2.1. 2-D plate of a periodic structure

The starting point of our investigation is the 2-D formulation of a plate problem for a plate with initial stresses; see Timoshenko and Woinowsky-Krieger (1959). In accordance with this model, the equilibrium equations of a plate with initial stresses can be written in the form,

$$\frac{\partial^2 M_{ij}}{\partial x_i \partial x_j} + \frac{\partial}{\partial x_i}[\sigma_{ij}^*(\mathbf{x}, \mathbf{x}/\varepsilon)\frac{\partial w^\varepsilon}{\partial x_j}] = f(\mathbf{x}, \mathbf{x}/\varepsilon) \quad \text{in } G. \tag{2.1.1}$$

The relationship between moments M_{ij} and curvatures $\dfrac{\partial^2 w^\varepsilon}{\partial x_m \partial x_n}$ is given by the formula,

$$M_{ij} = A_{ijmn}(\mathbf{x}/\varepsilon)\frac{\partial^2 w^\varepsilon}{\partial x_m \partial x_n}, \tag{2.1.2}$$

where

$$A_{1111} = A_{2222} = D(\mathbf{x}/\varepsilon),\ A_{1122} = A_{2211} = D(\mathbf{x}/\varepsilon)\nu(\mathbf{x}/\varepsilon), \tag{2.1.3}$$

$$A_{1212} = A_{2121} = D(\mathbf{x}/\varepsilon)[1 - \nu(\mathbf{x}/\varepsilon)],$$

are the moments calculated on the basis of the Kirchhoff-Love hypothesis; $D(\mathbf{x}/\varepsilon)=\dfrac{E(\mathbf{x}/\varepsilon)h^3(\mathbf{x}/\varepsilon)}{12[1-\nu^2(\mathbf{x}/\varepsilon)]}$; $E(\mathbf{x}/\varepsilon)$ and $\nu(\mathbf{x}/\varepsilon)$ are Young's modulus and Poisson's ratios; $h(\mathbf{x}/\varepsilon)$ is the local thickness of the plate; G is the 2-D region occupied by the plate; w^ε is the normal deflection of the plate; $\overset{*}{\sigma}_{ij}(\mathbf{x},\mathbf{x}/\varepsilon)$ are the initial in-plate stresses calculated on the basis of the 2-D elasticity model; $f(\mathbf{x},\mathbf{x}/\varepsilon)$ is the normal force applied to the plate (it is equal to the sum of the mass and surface forces applied to the plate); and the indexes i, j, m, n, and so on take values 1, 2 in this section.

Using the two-scale method (see Sect. 1.2), a function $Z(\mathbf{x},\mathbf{x}/\varepsilon)$ of the arguments $\mathbf{x}=(x_1,x_2)$ and $\mathbf{x}/\varepsilon$ is considered as a function $f(\mathbf{x},\mathbf{y})$ of "slow" variables $\mathbf{x}$ and "fast" variables $\mathbf{y}=\mathbf{x}/\varepsilon$. In accordance with this note, functions A_{ijmn} and $\overset{*}{\sigma}_{ij}$ will be written in the form $A_{ijmn}(\mathbf{y})$ and $\overset{*}{\sigma}_{ij}(\mathbf{x},\mathbf{y})$. Functions $A_{ijmn}(\mathbf{y})$ and $\overset{*}{\sigma}_{ij}(\mathbf{x},\mathbf{y})$ are periodic in $\mathbf{y}$ with periodicity cell Y corresponding to the period of the plate structure; see Fig. 2.1.

We assume that the plate is clamped along the surface ∂G (see Fig. 2.1). In this case, the boundary conditions can be written in the form

$$w^\varepsilon(\mathbf{x})=\frac{\partial w^\varepsilon}{\partial x_j}(\mathbf{x})n_j=0 \quad \text{on } \partial G, \tag{2.1.4}$$

where $\mathbf{n}$ is a normal to ∂G (see Fig. 2.1).

Asymptotic expansions and equations of equilibrium

We will study global deformations of the plate, as $\varepsilon\to 0$. To do this, we use the following asymptotic expansion: asymptotic expansion for deflection,

$$w^\varepsilon=w^{(0)}(\mathbf{x})+\varepsilon^2 w^{(2)}(\mathbf{x},\mathbf{y})+\ldots=w^{(0)}(\mathbf{x})+\sum_{k=2}^{\infty}\varepsilon^k w^{(k)}(\mathbf{x},\mathbf{y}), \tag{2.1.5}$$

asymptotic expansion for moments,

$$M_{ij}=\sum_{k=0}^{\infty}\varepsilon^k M_{ij}^{(k)}(\mathbf{x},\mathbf{y}). \tag{2.1.6}$$

Note that in expansion (2.1.5), the term $w^{(0)}(\mathbf{x})$ depends only on variables $\mathbf{x}$ and the term of the order of ε is absent.

Using the two-scale expansion method, the differential operators are presented in the form of the sum of operators in $\{x_i\}$ and in $\{y_i\}$ (see Sect. 1.2). For the function $Z(\mathbf{x},\mathbf{y})$ of the arguments $\mathbf{x}=(x_1,x_2)$ and $\mathbf{y}=(y_1,y_2)$, as on the right-hand sides of (2.1.5) and (2.1.6), this representation takes the form,

$$\frac{\partial Z}{\partial x_i}=Z_{,ix}+\varepsilon^{-1}Z_{,iy}, \tag{2.1.7}$$

$$\frac{\partial^2 Z}{\partial x_i \partial x_j}=Z_{,ixjx}+\varepsilon^{-1}(Z_{,ixjy}+Z_{,iyjx})+\varepsilon^{-2}Z_{,iyjy}\quad (i=1,2).$$

Substituting (2.1.5) and (2.1.6) in (2.1.1) and replacing the differential operators in accordance with the differentiating rule (2.1.7), we obtain

$$\sum_{k=0}^{\infty}\varepsilon^k[M^{(k)}_{ij,ixjx}+\varepsilon^{-1}(M^{(k)}_{ij,ixjy}+M^{(k)}_{ij,iyjx})+\varepsilon^{-2}M^{(k)}_{ij,iyjy}]+ \tag{2.1.8}$$

$$+\sum_{k=0}^{\infty}\varepsilon^k[\sigma^*_{ij}(w^{(k)}_{,jx}+\varepsilon^{-1}w^{(k)}_{,jy})]_{,ix}+\varepsilon^{-1}[\sigma^*_{ij}(w^{(k)}_{,jx}+\varepsilon^{-1}w^{(k)}_{,jy})]_{,iy}=0.$$

Denote

$$\langle\circ\rangle=(mes\,Y)^{-1}\int_Y \circ\,d\mathbf{y}$$

as the average value over periodicity cell $Y\subset R^2$, $\mathbf{y}=(y_1,y_2)$.

Averaging (2.1.8), we obtain

$$\sum_{k=0}^{\infty}\varepsilon^{k-2}\langle M^{(k)}_{ij}\rangle_{,ixjx}+\sum_{k=0}^{\infty}\varepsilon^k(\langle\sigma^*_{ij}\rangle w^{(k)}_{,jx})_{,ix}=\langle f\rangle(\mathbf{x}). \tag{2.1.9}$$

Here, we use the equality $\langle Z_{,jy}(\mathbf{y})\rangle=0$ which is fulfilled for every function $Z(\mathbf{y})$ periodic in $\mathbf{y}$. This equation follows from the equation,

$$\int_Y Z_{,jy}\,d\mathbf{y}=\int_{\partial Y} Zn_j\,d\mathbf{y},$$

the periodicity of function $Z(\mathbf{y})$, and the antiperiodicity of the normal vector $\mathbf{n}(\mathbf{y})$.

Proposition 2.1. *The following relationship holds: for* $\varepsilon\to 0$,

$$\int_G Z(\mathbf{x}, \mathbf{x}/\varepsilon)d\mathbf{x} = \int_G \langle Z\rangle(\mathbf{x})d\mathbf{x} \tag{2.1.10}$$

for every function $Z(\mathbf{x},\mathbf{y})$ *periodic in* $\mathbf{y}$ *with periodicity cell* Y. The proof can be found in Bensoussan et al. (1978).

Equating terms of the same order of ε in (2.1.9), we obtain, with allowance for Proposition 2.1, an infinite sequence of homogenized equations of equilibrium; the first has the following form:

$$\langle M_{ij}^{(0)}\rangle_{,ixjx} + (\langle \overset{*}{\sigma}_{ij}\rangle w_{,jx}^{(0)})_{,ix} = \langle f\rangle(\mathbf{x}). \tag{2.1.11}$$

Equating terms of the same order of ε in (2.1.8), we obtain an infinite sequence of local equations of equilibrium; the first has the following form

$$M_{ij,iyjy}^{(0)} = 0. \tag{2.1.12}$$

Substituting (2.1.5), (2.1.6) in (2.1.2) and replacing the differential operators in accordance with (2.1.7),

$$\sum_{k=0}^{\infty} \varepsilon^k M_{ij}^{(k)} \tag{2.1.13}$$

$$= \sum_{k=0}^{\infty} \varepsilon^k \{A_{ijmn}(\mathbf{y})[w_{,mxnx}^{(k)} + \varepsilon^{-1}(w_{,mxny}^{(k)} + w_{,mynx}^{(k)}) + \varepsilon^{-2} w_{,myny}^{(k)}]\}.$$

Equating terms of the same order of ε in (2.1.13), we obtain

$$M_{ij}^{(0)} = A_{ijmn}(\mathbf{y})w_{,myny}^{(2)} + A_{ijmn}(\mathbf{y})w_{,mxnx}^{(0)}(\mathbf{x}). \tag{2.1.14}$$

Substituting (2.1.14) in (2.1.12),

$$[A_{ijmn}(\mathbf{y})w_{m,ny}^{(2)} + A_{ijmn}(\mathbf{y})w_{,mxnx}^{(0)}(\mathbf{x})]_{,iyjy} = 0. \tag{2.1.15}$$

The Y-periodic solution of problem (2.1.15) is given by the formula

$$w^{(2)} = X^{kl}(\mathbf{y})w_{,kxlx}^{(0)}(\mathbf{x}) + V(\mathbf{x}), \tag{2.1.16}$$

where $X^{kl}(\mathbf{y})$ is the solution of the following cellular problem:

$$\begin{cases} [A_{ijmn}(\mathbf{y})X^{kl}_{,myny} + A_{ijkl}(\mathbf{y})]_{,iyjy} = 0 \quad \text{in } Y, \\ X^{kl}(\mathbf{y}) \text{ periodic in } \mathbf{y} \text{ with periodicity cell } Y. \end{cases} \tag{2.1.17}$$

Substituting (2.1.16) in (2.1.14), we obtain

$$M^{(0)}_{ij} = [A_{ijmn}(\mathbf{y})X^{kl}_{,myny}(\mathbf{y}) + A_{ijkl}(\mathbf{y})]w^{(0)}_{,kxlx}(\mathbf{x}) . \tag{2.1.18}$$

Averaging (2.1.18), we obtain the homogenized governing equation

$$\langle M^{(0)}_{ij} \rangle = D_{ijkl} w^{(0)}_{,kxlx}(\mathbf{x}) , \tag{2.1.19}$$

where

$$D_{ijkl} = \langle A_{ijmn}(\mathbf{y})X^{kl}_{,myny}(\mathbf{x}) + A_{ijkl}(\mathbf{y}) \rangle \tag{2.1.20}$$

are the homogenized stiffnesses of the plate.

Boundary conditions can be derived from the original boundary conditions (2.1.2) and asymptotic expansion (2.1.5). These are as follows:

$$w(\mathbf{x}) = \frac{\partial w}{\partial x_i}(\mathbf{x}) n_i = 0 \quad \text{on } \partial G . \tag{2.1.21}$$

Allowing for the homogenized governing equation (2.1.19), the homogenized equation of equilibrium (2.1.19) can be written as

$$\langle D_{ijkl} w_{,kxlx} \rangle_{,ixjx} + (\langle \sigma^{*}_{ij} \rangle w^{(0)}_{,jx})_{,ix} = \langle f \rangle(\mathbf{x}) . \tag{2.1.22}$$

Proposition 2.2. *Let the initial stresses* σ^{*}_{ij} *be determined from the solution of the* 2-D *elasticity problem. Then,*

$$\langle \sigma^{*}_{ij} \rangle = \sigma_{ij} . \tag{2.1.23}$$

Allowing for Proposition 2.2, we can write (2.1.22) as

$$\langle D_{ijkl} w_{,kxlx} \rangle_{,ixjx} + (\sigma_{ij} w^{(0)}_{,jx})_{,ix} = \langle f \rangle(\mathbf{x}) .$$

Dynamic problem for a stressed plate

The homogenized dynamic problem can be obtained as above, and it has the following form:

$$\langle M_{ij}^{(0)}\rangle_{,ixjx} + (\sigma_{ij}^{*} w_{,jx}^{(0)})_{,ix} = \langle \rho \rangle \frac{\partial^2 w^{(0)}}{\partial t^2} + \langle f \rangle(\mathbf{x},t), \qquad (2.1.24)$$

where $\rho(\mathbf{x})$ is the local density of the plate (it is equal to the specific density multiplied by the plate thickness).

The eigenfrequency problem has the form,

$$\langle M_{ij}^{(0)}\rangle_{,ixjx} + (\sigma_{ij}^{*} w_{,jx}^{(0)})_{,ix} = \omega^2 \langle \rho \rangle w^{(0)} + \langle f \rangle(\mathbf{x},t), \qquad (2.1.25)$$

where ω denotes an eigenfrequency.

2.2 Stressed Plates (Transition from 3-D to 2-D Model, In-Plane Initial Forces)

In this section, the transition from a 3-D elasticity problem with initial stresses to a plate theory problem is made on the basis of the homogenization method, and it is shown that the classical model is valid for plates of complex structure, which do not satisfy the Kirchhoff-Love hypothesis.

As will be seen, the asymptotic analysis for stressed plates has both similarities and differences in comparison with the analysis of a nonstressed plate given in the papers by Ciarlet and Destuyner (1979), Kohn and Vogelius (1984), Caillerie (1984), and Panasenko and Reztsov (1987) and with the analysis given in Chap. 1 for a stressed 3-D body. The differences are connected with the asymmetry of the coefficients, requiring a more detailed analysis of the problem. The difference between the problem of a stressed plate and the problem of a stressed 3-D body is connected with the difference in asymptotic expansions for a 3-D body and for a plate.

Formulation of the problem

We will examine a body of periodic structure obtained by repeating a certain small periodicity cell εY over the Ox_1x_2 plane (Fig. 2.2).

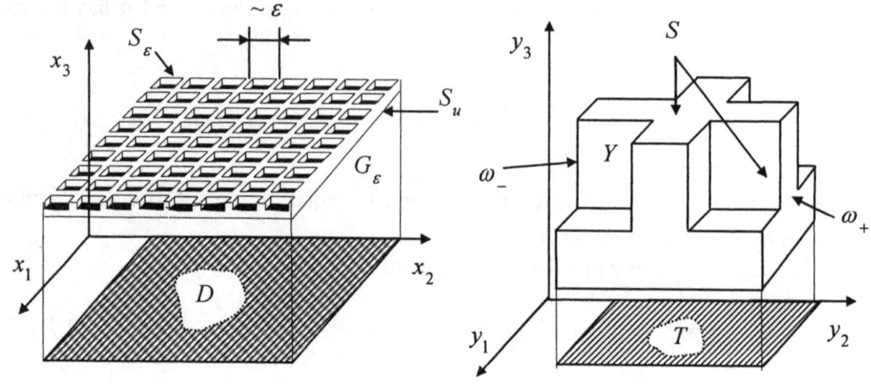

Fig. 2.2. 3-D plate of a periodic structure and its periodicity cell in fast variables

Here ε is the periodicity cell's characteristic dimension, which is assumed to be small (that is formalized in the form $\varepsilon \to 0$). As a result, we have a body of periodic structure with a small thickness - 3-D plate. For $\varepsilon \to 0$, the 3-D plate "tightens" to the domain D at the Ox_1x_2 plane (Fig. 2.2) - 2-D plate. Our aim is to derive a model describing the 2-D plate.

The starting point of our investigation is the exact 3-D formulation of an elasticity problem for the body with initial stresses without any simplifying assumptions, presented in Sect. 1.1. In accordance with this model, the equilibrium equations of the plate as a 3-D elastic body with initial stresses can be written in the form,

$$\frac{\partial}{\partial x_j}[h_{ijmn}(x_1,x_2,\mathbf{x}/\varepsilon)\frac{\partial u_m^\varepsilon}{\partial x_n}] + f_i^\varepsilon(\mathbf{x}) = 0 \quad \text{in } G_\varepsilon \,, \tag{2.2.1}$$

$$\mathbf{f}^\varepsilon(\mathbf{x}) = (\varepsilon^{-2} f_1, \varepsilon^{-2} f_2, \varepsilon^{-1} f_3)\,(x_1,x_2,\mathbf{x}/\varepsilon).$$

The boundary conditions can be written in the form,

$$h_{ijmn}(x_1,x_2,\mathbf{x}/\varepsilon)\frac{\partial u_m^\varepsilon}{\partial x_n} n_j = g_i^\varepsilon(\mathbf{x}/\varepsilon) \quad \text{on } S_\varepsilon \,, \tag{2.2.2}$$

$$\mathbf{g}^\varepsilon(\mathbf{x}) = (\varepsilon^{-1} g_1, \varepsilon^{-1} g_2, g_3)(x_1,x_2,\mathbf{x}/\varepsilon)\,,$$

$$\mathbf{u}^\varepsilon(\mathbf{x}) = \mathbf{u}^0(\mathbf{x}) \text{ on } S_u \,.$$

Here G_ε is the region occupied by the plate; S_ε is the lateral surface of the body, $\mathbf{n}$ is a normal to S_ε; S_u is the lateral surface along which the body is clamped (Fig. 2.2); $\mathbf{u}^\varepsilon$ are the displacements; $\mathbf{f}^\varepsilon$ and $\mathbf{g}^\varepsilon$ are the mass and surface forces applied to the body; h_{ijmn} are known (see Sect. 1.1.) combinations of the tensor of elasticity constants $\varepsilon^{-3}c_{ijmn}$ and the initial stresses $\overset{*}{\sigma}_{ij}$, which, in the case under consideration, are taken in the form

$$h_{ijmn}(x_1,x_2,\mathbf{x}/\varepsilon)=\varepsilon^{-3}c_{ijmn}(\mathbf{x}/\varepsilon)+\varepsilon^{-2}b^{(-2)}_{ijmn}(x_1,x_2,\mathbf{x}/\varepsilon) \tag{2.2.3}$$

$$+\varepsilon^{-1}b^{(-1)}_{ijmn}(x_1,x_2,\mathbf{x}/\varepsilon),$$

where

$$b^{(p)}_{ijmn}(x_1,x_2,\mathbf{x}/\varepsilon)=\overset{*}{\sigma}{}^{(p)}_{jn}(x_1,x_2,\mathbf{x}/\varepsilon)\delta_{im}\ (p=-2,-1). \tag{2.2.4}$$

Using the two-scale method, a function $Z(x_1,x_2,\mathbf{x}/\varepsilon)$ of the arguments x_1,x_2 and $\mathbf{x}/\varepsilon$ is considered a function $Z(x_1,x_2,\mathbf{y})$ of "slow" variables x_1,x_2 and "fast" variables $\mathbf{y}=\mathbf{x}/\varepsilon$. In accordance with this note, the functions c_{ijmn}, $\overset{*}{\sigma}_{jn}$ will be written in the form $c_{ijmn}(\mathbf{y})$, $\overset{*}{\sigma}_{jn}(x_1,x_2,\mathbf{y})$. The functions $c_{ijmn}(\mathbf{y})$, $\overset{*}{\sigma}_{jn}(x_1,x_2,\mathbf{y})$ are periodic in y_1,y_2 with periodicity cell T corresponding to the period of the plate structure (T is the projection of periodicity cell Y in "fast" variables $\mathbf{y}$ on the Oy_1y_2 plane; see Fig. 2.2).

For $h_{ijmn}(x_1,x_2,\mathbf{x}/\varepsilon)$ determined by (2.2.3), the relationship between additional stresses σ_{ij} and displacements $\mathbf{u}^\varepsilon$ takes the form,

$$\sigma_{ij}=h_{ijmn}(x_1,x_2,\mathbf{x}/\varepsilon)\frac{\partial u^\varepsilon_m}{\partial x_n}. \tag{2.2.5}$$

Formula (2.2.5) can be considered a local governing equation for a plate with initial stresses.

In connection with the fact that the coefficients (2.2.3) are written in a form different from that normally used, we will comment briefly on the terms in (2.3.3). The tensor $\varepsilon^{-3}c_{ijmn}$ describes the elastic constants of the material from which the plate is made. The multiplier ε^{-3} guarantees that the bending stiffnesses of the beam will be nonzero, as $\varepsilon\to 0$. The other terms describe the initial stresses.

The term $\varepsilon^{-1} b^{(-1)}_{ijmn} = \varepsilon^{-1} \overset{*}{\sigma}{}^{(-1)}_{jn} \delta_{im}$ corresponds to the resultant in-plane forces, which are independent of ε. In fact, the resultant forces are equal to the stress multiplied by the thickness, which has a characteristic value equal to ε. In classical theory, the plate buckles under the in-plane forces (corresponding to stresses $\varepsilon^{-1} \overset{*}{\sigma}{}^{(-1)}_{\alpha\beta}$). If the in-plane forces are zero, the local initial stresses in the 3-D plate can be nonzero. Can these stresses influence the behavior (in particular, the buckling) of the plate? In order to consider this problem, we introduce the term $\varepsilon^{-2} b^{(-2)}_{ijmn} = \varepsilon^{-2} \overset{*}{\sigma}{}^{(-2)}_{jn} \delta_{im}$ corresponding to the resultant moments. The condition that the tension force corresponding to $\overset{*}{\sigma}{}^{(-2)}_{jn}$ is zero, is written in the form,

$$\langle \overset{*}{\sigma}{}^{(-2)}_{\alpha\beta} \rangle = 0 \ (\alpha, \beta = 1, 2)\,. \tag{2.2.6}$$

We use the notations

$$\langle \circ \rangle = (mesT)^{-1} \int_Y \circ \, d\mathbf{y} \,, \text{ and } \langle \circ \rangle_S = (mesT)^{-1} \int_S \circ \, d\mathbf{y}$$

for the average value over periodicity cell Y and the lateral surface S in "fast" variables $\mathbf{y} = \mathbf{x} / \varepsilon$ (Fig. 2.2).

Furthermore in this section, b_{ijmn} means $b^{(-1)}_{ijmn}$, and $\overset{*}{\sigma}_{ij}$ means $\overset{*}{\sigma}{}^{(-1)}_{ij}$.

Asymptotic expansions and equations of equilibrium

We will study global deformations of the plate, as $\varepsilon \to 0$. To do this, we use the following asymptotic expansion used in the work by Caillerie (1984): asymptotic expansion for displacements,

$$\mathbf{u}^{\varepsilon} = \mathbf{u}^{(0)}(x_1, x_2) + \varepsilon \mathbf{u}^{(1)}(x_1, x_2, \mathbf{y}) + \ldots \tag{2.2.7}$$

$$= \mathbf{u}^{(0)}(x_1, x_2) + \sum_{k=1}^{\infty} \varepsilon^k \mathbf{u}^{(k)}(x_1, x_2, \mathbf{y})\,,$$

asymptotic expansion for stresses,

$$\sigma_{ij} = \sum_{p=-3}^{\infty} \varepsilon^p \sigma^{(p)}_{ij}(x_1, x_2, \mathbf{y})\,. \tag{2.2.8}$$

The functions on the right-hand side of (2.2.7) and (2.2.8) are assumed to be periodic in y_1, y_2 with periodicity cell T. Note that the term $\mathbf{u}^{(0)}(x_1, x_2)$ in (2.2.7) depends only on "slow" variable x_1, x_2. Expansion (2.2.8) starts with a term of order ε^{-3} in accordance with governing equation (2.2.5) and expansion (2.2.7) for displacements.

Using the two-scale expansion, the differential operators are represented in the form of the sum of operators in $\{x_i\}$ and in $\{y_i\}$. For the function $Z(x_1, x_2, \mathbf{y})$ of the arguments x_1, x_2 and $\mathbf{y} = (y_1, y_2, y_3)$, on the right-hand sides of (2.2.7) and (2.2.8), this representation takes the form,

$$\frac{\partial Z}{\partial x_3} = \varepsilon^{-1} Z_{,3y}, \tag{2.2.9}$$

$$\frac{\partial Z}{\partial x_\alpha} = Z_{,\alpha x} + \varepsilon^{-1} Z_{,\alpha y} \quad (\alpha = 1, 2).$$

Here and below in this chapter, the Latin indexes take the values 1, 2, 3; the Greek indexes take the values 1, 2; subscript $,\alpha x$ means $\partial/\partial x_\alpha$ $(\alpha = 1, 2)$, and subscript $,3y$ means $\partial/\partial x_3$.

The analysis of problem (2.2.1)–(2.2.4) breaks down into two stages. The first entails obtaining the equations of equilibrium for the plate considered as a 2-D structure. This stage is not involved with local governing equations [in the case under consideration, with (2.2.5)], and it is the same stage for any governing equations.

Proposition 2.3. *The following relationship holds: for* $\varepsilon \to 0$,

$$\varepsilon^{-1} \int_{G_\varepsilon} Z(x_1, x_2, \mathbf{x}/\varepsilon) \mathrm{d}\mathbf{x} \to \int_D \langle Z \rangle (x_1, x_2) \mathrm{d}x_1 \mathrm{d}x_2,$$

$$\int_{S_\varepsilon} Z(x_1, x_2, \mathbf{x}/\varepsilon) \mathrm{d}\mathbf{x} \to \int_D \langle Z \rangle (x_1, x_2) \mathrm{d}x_1 \mathrm{d}x_2.$$

The proof can be found in Caillerie (1984).

With allowance for (2.2.5), problem (2.2.1), (2.2.2) can be written as

$$\frac{\partial \sigma_{ij}}{\partial x_j} + f_i^\varepsilon(\mathbf{x}) = 0 \quad \text{in } G_\varepsilon, \quad \sigma_{ij} n_j = g_i^\varepsilon(\mathbf{x}/\varepsilon) \quad \text{on } S_\varepsilon. \tag{2.2.10}$$

To obtain the homogenized equilibrium equations, consider a weak form of problem (2.2.10), which has the form (see Sect. 1.3),

$$\int_{G_\varepsilon} \sigma_{ij} \frac{\partial v_i}{\partial x_j} d\mathbf{x} = \int_{S_\varepsilon} \mathbf{g}^\varepsilon \mathbf{v} d\mathbf{x} + \int_{G_\varepsilon} \mathbf{f}^\varepsilon \mathbf{v} d\mathbf{x} \quad \text{for any } \mathbf{v} \in V_\varepsilon . \tag{2.2.11}$$

Here, $\mathbf{v} \in V_\varepsilon$ is the trial function, and

$$V_\varepsilon = \{\mathbf{v} \in \{H^1(G_\varepsilon)\}^3 \;:\; \mathbf{v}(\mathbf{x}) = 0 \;\text{ on } S_u\} . \tag{2.2.12}$$

is the set of admissible displacements corresponding to the boundary conditions (2.2.2).

Substituting (2.2.8) in equilibrium equations (2.2.11), we obtain, with allowance for (2.2.9),

$$\sum_{p=-3}^{\infty} \int_{G_\varepsilon} \varepsilon^p \sigma_{ij}^{(p)} (v_{i,\alpha x} \delta_{j\alpha} + \varepsilon^{-1} v_{i,jy}) d\mathbf{x} = \int_{S_\varepsilon} \mathbf{g}^\varepsilon \mathbf{v} d\mathbf{x} + \int_{G_\varepsilon} \mathbf{f}^\varepsilon \mathbf{v} d\mathbf{x} \tag{2.2.13}$$

$$\text{for any } \mathbf{v} \in V_\varepsilon .$$

We introduce the functional space

$$U = \{\mathbf{w}(x_1, x_2) \in \{H^1(D)\}^3 \;:\; \mathbf{w}(x_1, x_2) = 0 \text{ on } \partial D\} . \tag{2.2.14}$$

It is obvious that $U \subset V_\varepsilon$. We take the trial function in the form $\mathbf{v} = \mathbf{w}(x_1, x_2) \in U$. Then, (2.2.13) takes the following form:

$$\sum_{p=-3}^{\infty} \int_{G_\varepsilon} \varepsilon^p \sigma_{i\alpha}^{(p)} w_{i,\alpha x} d\mathbf{x} = \int_{S_\varepsilon} \mathbf{g}^\varepsilon \mathbf{w} d\mathbf{x} + \int_{G_\varepsilon} \mathbf{f}^\varepsilon \mathbf{w} d\mathbf{x} \quad \text{for any } \mathbf{w} \in U .$$

Taking into account Proposition 2.3 and the definition of mass force $\mathbf{f}^\varepsilon$ (2.2.1), we obtain the following equality, as $\varepsilon \to 0$:

$$\sum_{p=-3}^{\infty} \int_D \varepsilon^{p+1} \langle \sigma_{i\alpha}^{(p)} \rangle w_{i,\alpha x} dx_1 dx_2 \tag{2.2.15}$$

$$= \int_D (\varepsilon^{-1} \langle g_\alpha \rangle_S w_\alpha + \langle g_3 \rangle_S w_3) dx_1 dx_2$$

$$+ \int_D (\varepsilon^{-1} \langle f_\alpha \rangle w_\alpha + \langle f_3 \rangle w_3) dx_1 dx_2 \;\text{ for any } \mathbf{w} \in U .$$

Integrating (2.2.15) by parts, we obtain

$$\sum_{p=-3}^{\infty} -\int_D \varepsilon^{p+1} \langle \sigma_{i\alpha}^{(p)} \rangle_{,\alpha x} w_i dx_1 dx_2 \tag{2.2.16}$$

$$= \int_D (\varepsilon^{-1} \langle g_\alpha \rangle_S w_\alpha + \langle g_3 \rangle_S w_3) \mathrm{d}x_1 \mathrm{d}x_2$$

$$+ \int_D (\varepsilon^{-1} \langle f_\alpha \rangle w_\alpha + \langle f_3 \rangle w_3) \mathrm{d}x_1 \mathrm{d}x_2 \quad \text{for any } \mathbf{w} \in U .$$

Introduce the resultant in-plane stresses by the formula (note that the integration in the formula is over 3-D cell Y, not over a plate thickness),

$$N_{ij}^{(p)}(x_1, x_2) = \langle \sigma_{ij}^{(p)}(x_1, x_2, \mathbf{y}) \rangle .$$

From (2.2.16), we obtain an infinite sequence of equilibrium equations for the resultant in-plane forces $N_{ij}^{(p)}$; the first three of them are the following (the others are not used) :

$$N_{\alpha\beta,\beta x}^{(-3)} = 0 \ , \tag{2.2.17}$$

$$N_{\alpha\beta,\beta x}^{(-2)} + \langle f_\alpha \rangle + \langle g_\alpha \rangle_S = 0,$$

$$N_{\alpha\beta,\beta x}^{(-1)} + \langle f_3 \rangle + \langle g_3 \rangle_S = 0 \tag{2.2.18}$$

($i, j = 1, 2, 3$ and $\alpha, \beta = 2, 3$).

Now, we take the trial function in the form $\mathbf{v} = \mathbf{w}(x_1, x_2) y_3$. It is obvious that $\mathbf{w}(x_1, x_2) y_3 \in V_\varepsilon$. Then, (2.2.13) takes the following form:

$$\sum_{p=-3}^{\infty} \int_{G_\varepsilon} \varepsilon^{p+1} \sigma_{ij}^{(p)} (w_{i,\alpha x} \delta_{j\alpha} y_3 + \varepsilon^{-1} w_i \delta_{j3}) \mathbf{dx}$$

$$= \int_{S_\sigma} \mathbf{g}^\varepsilon \mathbf{w} y_3 \mathrm{d}\mathbf{x} + \int_{G_\varepsilon} \mathbf{f}^\varepsilon \mathbf{w} y_3 \mathrm{d}\mathbf{x} \quad \text{for any } \mathbf{w} \in U .$$

With regard to Proposition 2.3, we obtain from this equation

$$\sum_{p=-3}^{\infty} \int_D \varepsilon^p (\langle \sigma_{i\alpha}^{(p)} y_3 \rangle w_{i,\alpha x} + \varepsilon^{-1} \langle \sigma_{i3}^{(p)} \rangle w_i) \mathrm{d}x_1 \mathrm{d}x_2 \tag{2.2.19}$$

$$= \int_D (\varepsilon^{-1} \langle g_\alpha y_3 \rangle_S w_\alpha + \langle g_3 y_3 \rangle_S w_3) \mathrm{d}x_1 \mathrm{d}x_2$$

$$+ \int_D (\varepsilon^{-1} \langle f_\alpha y_3 \rangle w_\alpha + \langle f_3 y_3 \rangle w_3) \mathrm{d}x_1 \mathrm{d}x_2 \quad \text{for any } \mathbf{w} \in U .$$

Integrating (2.2.19) by parts, we have

$$\sum_{p=-3}^{\infty} \int_D \varepsilon^{p+1} (-\langle \sigma_{i\alpha}^{(p)} y_3 \rangle_{,\alpha x} + \varepsilon^{-1} \langle \sigma_{i3}^{(p)} \rangle) w_i \mathrm{d}x_1 \mathrm{d}x_2 \tag{2.2.20}$$

$$= \int_D (\varepsilon^{-1} \langle g_\alpha y_3 \rangle_S w_\alpha + \langle g_3 y_3 \rangle_S w_3) \mathrm{d}x_1 \mathrm{d}x_2$$

$$+ \int_D (\varepsilon^{-1} \langle f_\alpha y_3 \rangle w_\alpha + \langle f_3 y_3 \rangle w_3) \mathrm{d}x_1 \mathrm{d}x_2 \quad \text{for any } \mathbf{w} \in U .$$

We introduce the resultant moments by the formula (note that the integration in the formula is over 3-D cell Y, not over a plate thickness)

$$M_{\alpha\beta}^{(p)}(x_1, x_2) = -\langle \sigma_{\alpha\beta}^{(p)}(x_1, x_2, \mathbf{y}) y_3 \rangle .$$

Note. Integration over a 3-D periodicity cell (not over a cross section) in the definition of the resultant in-plane forces and moments is related to the fact that we cannot determine the cross section of a plate of complex structure in a good way. If a plate has constant thickness and it is made of homogeneous material, then we can take $Y = [0,1]^3$, $T = [0,1]^2$. In the case under consideration, the local stresses have the form $\sigma_{ij}(x_1, x_2, \mathbf{y}) = \sigma_{ij}(x_1, x_2, y_3)$. As a result,

$$\langle \sigma_{ij}(x_1, x_2, \mathbf{y}) y_3^{\nu} \rangle = (mes T)^{-1} \int_Y \sigma_{ij}(x_1, x_2, y_3) y_3^{\nu} dx_1 dx_2 dx_3$$

$$= \int_0^1 \sigma_{ij}(x_1, x_2, y_3) y_3^{\nu} dx_3 \quad (\nu = 0,\ 1).$$

This means that for a plate of uniform structure made of homogeneous material, the definitions of the resultant in-plane forces and moments are equivalent to the classical definitions of resultant forces and moments.

From (2.2.20), we obtain the following equilibrium equations for the moments:

$$N_{i3}^{(-3)} = 0\,, \tag{2.2.21}$$

$$M_{\alpha\beta,\beta x}^{(-3)} - N_{\beta 3}^{(-2)} = 0\,, \tag{2.2.22}$$

$$M_{\alpha\beta,\beta x}^{(-2)} - N_{\beta 3}^{(-1)} + \langle f_\alpha y_3 \rangle + \langle g_\alpha y_3 \rangle_S = 0$$

$$(i, j = 1,\ 2,\ 3 \text{ and } \alpha, \beta = 2, 3).$$

We will not use the equilibrium equations corresponding to $p > -2$.

Substituting (2.2.8) in (2.2.10), we obtain, with allowance for the differentiating rule (2.2.9),

$$\sum_{p=-3}^{\infty} (\varepsilon^p \sigma_{i\alpha,\alpha x}^{(p)} + \varepsilon^{p-1} \sigma_{ij,jy}^{(p)}) + f_i^{\varepsilon}(\mathbf{x}) = 0 \quad \text{in } G_\varepsilon\ , \tag{2.2.23}$$

$$\sum_{p=-3}^{\infty} \varepsilon^p \sigma_{ij}^{(p)} n_j = g_i^{\varepsilon}(\mathbf{x}) \quad \text{on } S_\varepsilon\,.$$

Equating the terms with identical power in (2.2.23), we obtain the following sequence of local equations:

$$\sigma_{ij,jy}^{(p)} + \sigma_{i\alpha,\alpha x}^{(p-1)} = 0 \quad \text{in } Y\,, \tag{2.2.24}$$

$$\sigma_{ij}^{(p)} n_j = 0 \quad \text{on } S \ (p = -3, -2, \ldots),$$

$$\sigma_{ij}^{(-4)} = 0\,.$$

The governing equations

The second stage of analysis of the problem consists of obtaining the governing equations for the plate as a 2-D structure and excluding unknown quantities from the equilibrium equations. In contrast to the first stage, this stage does involve local governing equations.

Substituting (2.2.7) and (2.2.8) in the local governing equation (2.2.5), we obtain, with allowance for (2.2.9),

$$\sum_{p=-3}^{\infty} \varepsilon^{p} \sigma_{ij}^{(p)} \tag{2.2.25}$$

$$= \sum_{k=0}^{\infty} \varepsilon^{k} [\varepsilon^{-3} c_{ijmn}(\mathbf{y}) + \varepsilon^{-1} b_{ijmn}(x_1, x_2, \mathbf{y})](u_{m,\beta x}^{(k)} \delta_{m\beta} + \varepsilon^{-1} u_{m,ny}^{(k)}) .$$

Equating the terms with identical power of ε in (2.2.25), we obtain an infinite sequence of problems; the first three have the following form:

$$\sigma_{ij}^{(p)} = c_{ijm\beta}(\mathbf{y}) u_{m,\beta x}^{(p+3)} + c_{ijmn}(\mathbf{y}) u_{m,ny}^{(p+4)} \quad (p=-3,-2), \tag{2.2.26}$$

$$\sigma_{ij}^{(-1)} = c_{ijm\beta}(\mathbf{y}) u_{m,\beta x}^{(2)} + c_{ijmn}(\mathbf{y}) u_{m,ny}^{(3)} \tag{2.2.27}$$

$$+ b_{ijm\beta}(x_1, x_2, \mathbf{y}) u_{m,\beta x}^{(0)} + b_{ijmn}(x_1, x_2, \mathbf{y}) u_{m,ny}^{(1)} .$$

As was assumed above,

$$\mathbf{u}^{(k)}(\mathbf{y}) \text{ is periodic in } y_1, y_2 \text{ with periodicity cell } T, \quad k=1, 2, \ldots . \tag{2.2.28}$$

Let us consider the problem (2.2.24, $p=-3$), (2.2.26, $p=-3$), and (2.2.28, $k=1$). It can be written in the form,

$$\begin{cases} [c_{ijmn}(\mathbf{y}) u_{m,ny}^{(1)} + c_{ijm\alpha}(\mathbf{y}) u_{m,\alpha x}^{(0)}(x_1, x_2)]_{,jy} = 0 \quad \text{in } Y, \\ [c_{ijmn}(\mathbf{y}) u_{m,ny}^{(1)} + c_{ijm\alpha}(\mathbf{y}) u_{m,\alpha x}^{(0)}(x_1, x_2)] n_j = 0 \quad \text{on S}, \\ \mathbf{u}^{(1)}(\mathbf{y}) \text{ is periodic in } y_1, y_2 \text{ with periodicity cell } T. \end{cases} \tag{2.2.29}$$

Allowing for the fact that the function of argument x_1, x_2 plays the role of a parameter in the problems in the variables $\mathbf{y}$, solution of the problem (2.2.29) can be found in the form,

$$\mathbf{u}^{(1)} = \mathbf{X}^{0m\alpha}(\mathbf{y})u^{(0)}_{m,\alpha x}(x_1, x_2) + \mathbf{V}(x_1, x_2)\,. \tag{2.2.30}$$

Here, $\mathbf{V}(x_1, x_2)$ is an arbitrary function of argument x_1, x_2 (it will be determined below), and $\mathbf{X}^{0m\alpha}(\mathbf{y})$ is the solution of the following cellular problem:

$$\begin{cases} [c_{ijmn}(\mathbf{y})X^{0m\alpha}_{m,ny} + c_{ijm\alpha}(\mathbf{y})]_{,jy} = 0 \quad \text{in } Y, \\ [c_{ijmn}(\mathbf{y})X^{0m\alpha}_{m,ny} + c_{ijm\alpha}(\mathbf{y})]n_j = 0 \quad \text{on } S, \\ \mathbf{X}^{0m\alpha}(\mathbf{y}) \text{ is periodic in } y_1, y_2 \text{ with periodicity cell } T \end{cases} \tag{2.2.31}$$

$(\alpha, \beta = 1, 2$ and m=1, 2, 3).

Here, $\mathbf{n}$ means a normal to S (Fig. 2.2).

Proposition 2.4. *For m=3, the solution of cellular problem* (2.2.31) *is given by the following formula*:

$$\mathbf{X}^{03\alpha}(\mathbf{y}) = -y_3\mathbf{e}_\alpha\,. \tag{2.2.32}$$

Here, $\{\mathbf{e}_i\}$ are basis vectors of the coordinate system.

One can verify formula (2.2.32) by substituting (2.2.32) in cellular problem (2.2.31) and taking into account that $c_{ij3\alpha} = c_{ij\alpha 3}$; see condition *C1 a*), Sect. 1.2.

Taking into account equality (2.2.32), we can write (2.2.30) as

$$\mathbf{u}^{(1)} = \mathbf{X}^{0\beta\alpha}(\mathbf{y})u^{(0)}_{\beta,\alpha x}(x_1, x_2) - y_3\mathbf{e}_\alpha u^{(0)}_{3,\alpha x}(x_1, x_2) + \mathbf{V}(x_1, x_2)\,. \tag{2.2.33}$$

Substituting (2.2.33) in (2.2.26, p=–3) gives the following equality:

$$\sigma^{(-3)}_{ij} = c_{ij\beta\alpha}(\mathbf{y})u^{(0)}_{\beta,\alpha x}(x_1, x_2) + c_{ijkl}(\mathbf{y})X^{0\beta\alpha}_{k,ly}(\mathbf{y})u^{(0)}_{\beta,\alpha x}(x_1, x_2)\,. \tag{2.2.34}$$

Averaging (2.2.34) over periodicity cell Y with $ij = \gamma\delta$, we obtain

$$\langle \sigma^{(-3)}_{\gamma\delta} \rangle = A_{\gamma\delta\beta\alpha}u^{(0)}_{\beta,\alpha x}\,, \tag{2.2.35}$$

where

$$A_{\gamma\delta\alpha\beta} = \langle c_{\gamma\delta\alpha\beta}(\mathbf{y}) + c_{\gamma\delta kl}(\mathbf{y})X^{0\alpha\beta}_{k,ly}(\mathbf{y}) \rangle\,.$$

In accordance with the homogenized equilibrium equation (2.2.16, p=–3),

$$(A_{\gamma\delta\beta\alpha}u^{(0)}_{\beta,\alpha x})_{,\delta x} = 0 \text{ in } D. \tag{2.2.36}$$

The boundary condition for $u^{(0)}_\alpha$ ($\alpha = 1, 2$) follows from (2.2.2), and expansion (2.2.7) and has the form,

$$u^{(0)}_\alpha(x_1, x_2) = 0 \quad \text{on } \partial D \ (\alpha = 1, 2). \tag{2.2.37}$$

The solution of problem (2.2.36) and (2.2.37) is $u^{(0)}_\alpha(x_1, x_2) = 0$ ($\alpha = 1, 2$). Then,

$$\mathbf{u}^{(1)} = -y_3 \mathbf{e}_\alpha u^{(0)}_{3,\alpha x}(x_1, x_2) + \mathbf{V}(x_1, x_2), \tag{2.2.38}$$

and equalities (2.2.26, p=–3,–2) take the form,

$$\sigma^{(-3)}_{ij} = 0, \tag{2.2.39}$$

$$\sigma^{(-2)}_{ij} = c_{ijmn}(\mathbf{y})u^{(2)}_{m,ny} - c_{ij\alpha\beta}(\mathbf{y})y_3 u^{(0)}_{3,\alpha x\beta x}(x_1, x_2) \tag{2.2.40}$$

$$+ c_{ij\alpha\beta}(\mathbf{y})V_{\alpha,\beta x}(x_1, x_2).$$

Let us examine problem (2.2.24, p=–2), (2.2.40), and (2.2.28, k=2). It can be written in the form,

$$\begin{cases} [c_{ijmn}(\mathbf{y})u^{(2)}_{m,ny} - c_{ij\alpha\beta}(\mathbf{y})y_3 u^{(0)}_{3,\alpha x\beta x}(x_1, x_2) \\ \qquad + c_{ij\alpha\beta}(\mathbf{y})V_{\alpha,\beta x}(x_1, x_2)]_{,jy} = 0 \quad \text{in } Y, \\ [c_{ijmn}(\mathbf{y})u^{(2)}_{m,ny} - c_{ij\alpha\beta}(\mathbf{y})y_3 u^{(0)}_{3,\alpha x\beta x}(x_1, x_2) \\ \qquad + c_{ij\alpha\beta}(\mathbf{y})V_{\alpha,\beta x}(x_1, x_2)]n_j = 0 \quad \text{on } S, \\ \mathbf{u}^{(2)}(\mathbf{y}) \text{ is periodic in } y_1, y_2 \text{ with periodicity cell } T. \end{cases} \tag{2.2.41}$$

In order to solve problems of this kind, we introduce the cellular problem (according to bending of a plate)

$$\begin{cases} [c_{ijmn}(\mathbf{y})X^{1\alpha\beta}_{m,ny} - y_3 c_{ij\alpha\beta}(\mathbf{y})]_{,jy} = 0 \text{ in } Y, \\ [c_{ijmn}(\mathbf{y})X^{1\alpha\beta}_{m,ny} - y_3 c_{ij\alpha\beta}(\mathbf{y})]n_j = 0 \text{ on S}, \\ \mathbf{X}^{1\alpha\beta}(\mathbf{y}) \text{ is periodic in } y_1, y_2 \text{ with periodicity cell } T \end{cases} \tag{2.2.42}$$

($\nu = 0,1$; $\alpha, \beta = 1,2$). The solution of problem (2.2.41) can be expressed through the functions $\mathbf{X}^{\nu\alpha\beta}(\mathbf{y})$ - solutions of cellular problems (2.2.31) and (2.2.42) as follows:

$$\mathbf{u}^{(2)} = \mathbf{X}^{1\alpha\beta}(\mathbf{y})u^{(0)}_{3,\alpha x\beta x}(x_1,x_2) - y_3 V_{3,\beta x}(x_1,x_2)\mathbf{e}_\beta \tag{2.2.43}$$

$$+\mathbf{X}^{0\alpha\beta}(\mathbf{y})V_{\alpha,\beta x}(x_1,x_2).$$

Substituting (2.2.43) in (2.2.40),

$$\sigma^{(-2)}_{ij} = [c_{ijmn}(\mathbf{y})X^{0\alpha\beta}_{m,ny}(\mathbf{y}) + c_{ij\alpha\beta}(\mathbf{y})]V_{\alpha,\beta x}(x_1,x_2) \tag{2.2.44}$$

$$+[c_{ijmn}(\mathbf{y})X^{1\alpha\beta}_{m,ny}(\mathbf{y}) - y_3 c_{ij\alpha\beta}(\mathbf{y})]u^{(0)}_{3,\alpha x\beta x}(x_1,x_2).$$

Governing equations for resultant in-plane stresses and moments

Averaging (2.2.44) with $ij = \gamma\delta$ ($\gamma, \delta = 1,2$) over periodicity cell Y, we obtain

$$N^{(-2)}_{\gamma\delta} = A^0_{\gamma\delta\alpha\beta}V_{\alpha,\beta x} + A^1_{\gamma\delta\alpha\beta}u^{(0)}_{3,\alpha x\beta x}. \tag{2.2.45}$$

Multiplying (2.2.44) with $ij = \gamma\delta$ ($\gamma, \delta = 1,2$) by $-y_3$ and averaging over periodicity cell Y, we obtain the following equation:

$$M^{(-2)}_{\gamma\delta} = A^1_{\gamma\delta\alpha\beta}V_{\alpha,\beta x} + A^2_{\gamma\delta\alpha\beta}u^{(0)}_{3,\alpha x\beta x}. \tag{2.2.46}$$

The coefficients in (2.2.45) and (2.2.46) are

$$A^{\nu+\mu}_{\gamma\delta\alpha\beta} = \langle [(-1)^\nu c_{\gamma\delta\alpha\beta}(\mathbf{y})y_3^\nu + c_{\gamma\delta mn}(\mathbf{y})X^{\nu\alpha\beta}_{m,ny}(\mathbf{y})](-1)^\mu y_3^\mu \rangle \tag{2.2.47}$$

$$(\nu, \mu = 0,1;\ \alpha, \beta, \gamma, \delta = 1, 2).$$

The equations (2.2.45) and (2.2.46) obtained are governing equations of plate considered as a 2-D structure. The coefficients (2.2.47) are the homogenized stiffnesses of the plate. As we see, the stiffnesses are expressed through solutions of cellular problems (2.2.31) and (2.2.42). The governing equations derived determine the resultant in-plane forces and moments but not the shear forces.

Governing equations for shear forces

Collect the nontrivial homogenized equilibrium equations. These are the following [see (2.2.18) and the second equation form (2.2.22)]:

$$N^{(-2)}_{\alpha\beta,\beta x} + \langle f_\alpha \rangle + \langle g_\alpha \rangle_S = 0, \tag{2.2.48}$$

$$N^{(-1)}_{3\alpha,\alpha x} + \langle f_3 \rangle + \langle g_3 \rangle_S = 0, \tag{2.2.49}$$

$$M^{(-2)}_{\alpha\beta,\beta x} - N^{(-1)}_{\beta 3} + \langle f_\alpha y_3 \rangle + \langle g_\alpha y_3 \rangle_S = 0 \tag{2.2.50}$$

$$(i, j = 1,2,3;\ \alpha, \beta = 1,2).$$

In the absence of initial stresses, due to $\sigma^{(-1)}_{ij} = \sigma^{(-1)}_{ji}$, $N^{(-1)}_{ij} = N^{(-1)}_{ji}$. Then, we can exclude the quantities $N^{(-1)}_{3\alpha}$, $N^{(-1)}_{\alpha 3}$ (called the shear forces) from (2.2.49) and (2.2.50) using this symmetry with respect to the indexes i and j, only, as done in both the classical theory (see, e.g., Timoshenko and Woinowsky-Krieger, 1959) and in the asymptotic theory for plates with no initial stresses (see, e.g., Caillerie, 1984).

In our case $N^{(-1)}_{ij}$ is not symmetrical with respect to indexes i and j due to (2.2.27) [because b_{ijmn} are not symmetric in i and j, see definition of b_{ijmn} (2.2.4)]. As seen from (2.2.4), the asymmetry is directly related to the initial stresses. Then we need to examine $N^{(-1)}_{ij}$ in detail to obtain some additional information on them.

Let us insert $\mathbf{u}^{(0)}$ into (2.2.27) in accordance with (2.2.37). Then we obtain

$$\sigma^{(-1)}_{ij} = c_{ijm\beta}(\mathbf{y}) u^{(2)}_{m,\beta x} + c_{ijmn}(\mathbf{y}) u^{(3)}_{m,ny} \tag{2.2.51}$$

$$+ b_{ij3\beta}(x_1, x_2, \mathbf{y}) u^{(0)}_{3,\beta x}(x_1, x_2) - b_{ij\beta 3}(x_1, x_2, \mathbf{y}) u^{(0)}_{3,\beta x}(x_1, x_2).$$

The first and second terms on the right-hand side of equation (2.2.51) are symmetrical with respect to i and j by virtue of the symmetry of the elastic constants (see Sect. 1.2). Consider the asymmetrical part of the tensor $\sigma_{ij}^{(-1)}$. The following relation can be derived from (2.2.51):

$$\sigma_{ij}^{(-1)} - \sigma_{ji}^{(-1)} = b_{ij3\beta}(x_1, x_2, \mathbf{y})u_{3,\beta x}^{(0)} - b_{ji3\beta}(x_1, x_2, \mathbf{y})u_{3,\beta x}^{(0)} \tag{2.2.52}$$

$$- b_{ij\beta 3}(x_1, x_2, \mathbf{y})u_{3,\beta x}^{(0)} + b_{ji\beta 3}(x_1, x_2, \mathbf{y})u_{3,\beta x}^{(0)} .$$

Averaging (2.2.52) over periodicity cell Y, we obtain

$$N_{ij}^{(-1)} - N_{ji}^{(-1)} \tag{2.2.53}$$

$$= \langle b_{ij3\beta}(x_1, x_2, \mathbf{y})\rangle u_{3,\beta x}^{(0)}(x_1, x_2) - \langle b_{ji3\beta}(x_1, x_2, \mathbf{y})\rangle u_{3,\beta x}^{(0)}(x_1, x_2)$$

$$- \langle b_{ij\beta 3}(x_1, x_2, \mathbf{y})\rangle u_{3,\beta x}^{(0)}(x_1, x_2) + \langle b_{ji\beta 3}(x_1, x_2, \mathbf{y})\rangle u_{3,\beta x}^{(0)}(x_1, x_2) .$$

For $i = 3$ and $j = \gamma$, we obtain from (2.2.53) and the definition of b_{ijmn} (2.2.4),

$$N_{3\gamma}^{(-1)} - N_{\gamma 3}^{(-1)} = K_\gamma , \tag{2.2.54}$$

where $K_\gamma(x_1, x_2)$ is defined by the following formula:

$$K_\gamma(x_1, x_2) \tag{2.2.55}$$

$$= \langle b_{3\gamma 3\beta}(x_1, x_2, \mathbf{y})\rangle u_{3,\beta x}^{(0)}(x_1, x_2) - \langle b_{\gamma 33\beta}(x_1, x_2, \mathbf{y})\rangle u_{3,\beta x}^{(0)}(x_1, x_2)$$

$$- \langle b_{3\gamma\beta 3}(x_1, x_2, \mathbf{y})\rangle u_{3,\beta x}^{(0)}(x_1, x_2) - \langle b_{\gamma 3\beta 3}(x_1, x_2, \mathbf{y})\rangle u_{3,\beta x}^{(0)}(x_1, x_2)$$

$$= -\delta_{\gamma\beta}\langle \overset{*}{\sigma}_{33}(x_1, x_2, \mathbf{y})\rangle u_{3,\beta x}^{(0)}(x_1, x_2) + \langle \overset{*}{\sigma}_{\gamma\delta}(x_1, x_2, \mathbf{y})\rangle u_{3,\delta x}^{(0)}(x_1, x_2)$$

$$= -\overset{*}{N}_{33} u_{3,\gamma x}^{(0)}(x_1, x_2) + \overset{*}{N}_{\gamma\delta} u_{3,\delta x}^{(0)}(x_1, x_2) .$$

Here, we use the notation,

$$N^*_{ij} = \langle \sigma^*_{ij}(x_1, x_2, \mathbf{y}) \rangle \,, \tag{2.2.56}$$

for the average value of the initial stresses.

Proposition 2.5. *Let the initial stresses* σ^*_{ij} *satisfy the equilibrium equations,*

$$\frac{\partial \sigma^*_{ij}}{\partial x_j} + \varepsilon^a f_i(x_1, x_2, \mathbf{y}) = 0 \quad \text{in } G_\varepsilon \,, \tag{2.2.57}$$

$$\sigma^*_{ij} n_j = g_i(x_1, x_2, \mathbf{y}) \quad \text{on } S_\varepsilon \,.$$

If $a<-1$ *and* $\mathbf{g}=0$, *then,* $\langle b_{ijmn} Z_{,ny} \rangle = 0$ *for every differentiable function* $Z(\mathbf{y})$ *periodic in* y_1, y_2 *with periodicity cell* T .

Proof. Taking into account the definition of b_{ijmn} (2.2.4), it is sufficient prove that $\langle \sigma^*_{jn}(x_1, x_2, \mathbf{y}) Z_{,ny}(\mathbf{y}) \rangle$. From (2.2.57) with allowance for (2.2.9), we obtain

$$\sigma^*_{ij,jy} = F_i(x_1, x_2, \mathbf{y}) \quad \text{in } Y \ (F_i = 0 \text{ if } a \neq -1 \,,\ F_i = -f_i \text{ if } a = -1), \tag{2.2.58}$$

$$\sigma^*_{ij} n_j = g_i(x_1, x_2, \mathbf{y}) \quad \text{on } S \,,$$

$\sigma^*_{ij}(x_1, x_2, \mathbf{y})$ is periodic in y_1, y_2 with periodicity cell T .

Let us consider the quantity $\langle \sigma^*_{jn}(x_1, x_2, \mathbf{y}) Z_{,ny}(\mathbf{y}) \rangle$. Taking into account the definition of the average value over periodicity cell Y and integrating by parts, one can find that this quantity is equal to

$$-\int_Y \sigma^*_{jn,ny} Z(\mathbf{y}) d\mathbf{y} + \int_S \sigma^*_{jn} n_n Z(\mathbf{y}) d\mathbf{y} + \int_\omega \sigma^*_{j\gamma} n_\gamma Z(\mathbf{y}) d\mathbf{y} \tag{2.2.59}$$

$$= -\int_Y F_j Z(\mathbf{y}) d\mathbf{y} + \int_S g_j Z(\mathbf{y}) d\mathbf{y} \,.$$

Here, $\omega = \omega_+ \cup \omega_-$ means the opposite faces of periodicity cell Y normal to the Oy_1y_2 plane (Fig. 2.2). The integrals over Y and the right-hand side of equation (2.2.59) are equal to zero as a consequence of (2.2.58) (if $a \neq -1$ and $\mathbf{g} = 0$). The integral over ω is equal to zero as a consequence of the periodicity of the functions σ^*_{jn} and Z and the antiperiodicity of the vector normal to ω (Fig. 2.2). That proves the proposition.

Proposition 2.6. *Under the conditions of Proposition 2.5,* $\langle \sigma^*_{i3} \rangle = 0$.

Proposition 2.6 is a consequence of Proposition 2.5 (it is sufficient to set $Z(\mathbf{y}) = y_3$ into Proposition 2.5). From Proposition 2.6, it follows that $K_\gamma(x_1, x_2) = N^*_{\gamma\delta} u^{(0)}_{3,\delta x}(x_1, x_2)$,

The limit 2-D problem

Now, we can exclude the quantities $N^{(-1)}_{\beta 3}$, $N^{(-1)}_{3\beta}$ from equilibrium equations (2.2.49) and (2.2.50) and obtain a closed 2-D model of the plate. Differentiating (2.2.50) and using (2.2.49), (2.2.54), (2.2.55), and Proposition 2.6, we obtain

$$(A^0_{\gamma\delta\alpha\beta} V_{\alpha,\beta x} + A^1_{\gamma\delta\alpha\beta} u^{(0)}_{3,\alpha x\beta x})_{,\delta x} + \langle f_\gamma \rangle + \langle g_\gamma \rangle_S = 0, \tag{2.2.60}$$

$$(A^1_{\gamma\delta\alpha\beta} V_{\alpha,\beta x} + A^2_{\gamma\delta\alpha\beta} u^{(0)}_{3,\alpha x\beta x})_{,\gamma x\delta x} + \langle f_3 \rangle + \langle g_3 \rangle_S$$

$$+ (\langle f_\gamma y_3 \rangle + \langle g_\gamma y_3 \rangle_S)_{,\gamma x} = K_{\gamma,\gamma x} = (N^*_{\gamma\delta} u^{(0)}_{3,\delta x})_{,\gamma x}.$$

The closed 2-D model of a stressed plate consists of the equilibrium equation (2.2.48), (2.2.60), the governing equations (2.2.45), (2.2.48), and boundary conditions

$$V_\alpha(x_1, x_2) = 0 \ (\alpha = 2, 3), \tag{2.2.61}$$

$$u^{(0)}_3(x_1, x_2) = \frac{\partial u^{(0)}_3}{\partial x_i}(x_1, x_2) n_i = 0 \quad \text{on } \partial D.$$

Note (about the conditions from Proposition 2.5). Consider the conditions of Proposition 2.5. These conditions are satisfied for many cases. For example, condition $\mathbf{g}^\varepsilon = 0$ is satisfied if there are no surface forces applied to the plate. If the conditions of Proposition 2.5 are not satisfied then the nonclassical term $N^*_{33} u^{(0)}_{3,\gamma x}$

arises in the limit problem. We compute this term for the case $\mathbf{g}^{\varepsilon} \neq 0$ and $a = -1$. From (2.2.59) [with $Z(\mathbf{y}) = y_3$], integrating by parts, we obtain

$$\langle N^*_{33} \rangle = \langle \sigma^*_{33} \rangle = -\langle \sigma^*_{3n,ny} y_3 \rangle = -\int_Y f_3\, y_3 \mathbf{dy} - \int_S g_3 y_3 \mathbf{dy}.$$

For a plate with planar surfaces,

$$\langle N^*_{33} \rangle = \langle \sigma^*_{33} \rangle = -\int_Y f_3\, y_3 \mathbf{dy} - (g_3^+ - g_3^-)\frac{h}{2},$$

where g_3^+ and g_3^- are values of the surface force on the upper and lower faces of the plate and h is the plate thickness.

Dynamic problem

One can apply the homogenization method to a dynamic problem. Consider the case when in-plane dynamic effects may be neglected. The limit 2-D problem consists of (2.2.45), (2.2.46), (2.2.48), and (2.2.61) and the equation,

$$(A^1_{\gamma\delta\alpha\beta} V_{\alpha,\beta x} + A^2_{\gamma\delta\alpha\beta} u^{(0)}_{3,\alpha x\beta x})_{,\gamma x\delta x} + \langle f_3 \rangle + \langle g_3 \rangle = (N^*_{\gamma\delta} u^{(0)}_{3,\delta x})_{,\gamma x} + \langle \rho \rangle \frac{\partial^2 u_3^{(0)}}{\partial t^2}$$

for dynamic problem, or

$$(A^1_{\gamma\delta\alpha\beta} V_{\alpha,\beta x} + A^2_{\gamma\delta\alpha\beta} u^{(0)}_{3,\alpha x\beta x})_{,\gamma x\delta x} = (N^*_{\gamma\delta} u^{(0)}_{3,\delta x})_{,\gamma x} + \langle \rho \rangle \omega^2 u_3^{(0)}$$

for eigenfrequency problem.

Here $\rho(\mathbf{x}, \mathbf{x}/\varepsilon)$ means the local density, and $\langle \rho \rangle$ is the average density.

A uniform plate made of homogeneous material. Comparison with the classical case

Consider a uniform plate made of a homogeneous isotropic material. In this case, $c_{ijkl} = const$ and $Y = [0,1]^3$. In the case under consideration, we can find the solution of cellular problems (2.2.29) and (2.2.40) in the form $X_3^{\nu\alpha\beta} = X^{\nu\alpha\beta}(y_3)$ ($\nu = 0,1$), $X_\gamma^{\nu\alpha\beta} = 0$ ($\gamma = 1,2$). For these functions, cellular problems (2.2.29) and (2.2.40) can be written as

$$[c_{3333}X^{\nu\alpha\beta\,\prime}+c_{33\alpha\beta}(-1)^{\nu}y_3^{\nu}]'=0,\quad y_3\in[0,1], \tag{2.2.62}$$

$$c_{3333}X^{\nu\alpha\beta\,\prime}+c_{33\alpha\beta}(-1)^{\nu}y_3^{\nu}=0\quad\text{for } y_3=0,1,$$

where the prime means d/dy_3.

The function $X^{\nu\alpha\beta}(y_3)$ is periodic in y_1, y_2. The solution of problem (2.2.62) is $X^{\nu\alpha\beta\,\prime}(y_3)=(-1)^{\nu}y_3^{\nu}\dfrac{c_{33\alpha\beta}}{c_{3333}}$. In the case under consideration, (2.2.47) becomes

$$A^{\mu+\nu}_{\gamma\delta\alpha\beta}=\langle[(-1)^{\nu}c_{\gamma\delta\alpha\beta}y_3^{\nu}-c_{\gamma\delta 33}X^{\nu\alpha\beta\,\prime}(y_3)](-1)^{\mu}y_3^{\mu}\rangle. \tag{2.2.63}$$

Substituting (2.2.62) in (2.2.63), we obtain

$$A^{\mu+\nu}_{\gamma\delta\alpha\beta}=\langle[c_{\gamma\delta\alpha\beta}-\frac{c_{\gamma\delta 33}c_{33\alpha\beta}}{c_{3333}}(y_3)](-1)^{\mu+\nu}y_3^{\mu+\nu}\rangle \tag{2.2.64}$$

$$=\left(c_{\gamma\delta\alpha\beta}-\frac{c_{\gamma\delta 33}c_{33\alpha\beta}}{c_{3333}}\right)\langle(-y_3)^{\nu+\mu}\rangle.$$

The elastic constant tensor for an isotropic homogeneous material has the form,

$$c_{ijkl}=\frac{E\nu}{(1+\nu)(1-2\nu)}\delta_{ij}\delta_{kl}+\frac{E}{2(1+\nu)}(\delta_{ik}\delta_{jl}+\delta_{lj}\delta_{jk}).$$

In particular,

$$c_{1111}=c_{3333}=\frac{E(1-\nu)}{(1+\nu)(1-2\nu)},\quad c_{1133}=\frac{E\nu}{(1+\nu)(1-2\nu)}. \tag{2.2.65}$$

From (2.2.65),

$$c_{1111}-\frac{c_{1133}^2}{c_{3333}}=\frac{1}{1-\nu^2},$$

and we obtain the following formula for stiffnesses $A^{\mu+\nu}_{1111}$:

$$A^{\mu+\nu}_{1111}=\frac{E}{1-\nu^2}\langle(-y_3)^{\nu+\mu}\rangle.$$

It can be written as

$$A_{1111}^{0}=\frac{Eh}{1-\nu^{2}}\,,\ A_{1111}^{1}=0\,,\ A_{1111}^{2}=\frac{Eh^{3}}{12(1-\nu^{2})}\,,$$

where h is the thickness of the plate.

Thus, the 2-D model of a stressed plate, obtained using the homogenization method, coincides both qualitatively and quantitatively with the classical equations for a uniform plate made of a homogeneous material.

2.3 Stressed Plates (Transition from 3-D to 2-D Model, Moments of Initial Stresses)

In this section, we demonstrate that the initial stresses corresponding to bending/torsional moments do not influence the homogenized model.

Formulation of the problem

Now, we consider problem (2.2.1), (2.2.2) for the case $\{\sigma_{ij}^{*(-2)}\neq 0\,,\ \sigma_{ij}^{*(-1)}=0\}$, where $\sigma_{ij}^{*(-2)}$ satisfies condition (2.2.6). This means that the initial stresses generate no resultant stresses but nonzero resultant moments.

The starting point of our investigation is the exact 3-D formulation of elasticity problem (2.2.1) and (2.2.2).

In the case under consideration, the coefficients h_{ijmn} are the following:

$$h_{ijmn}(x_1,x_2,\mathbf{x}/\varepsilon)=\varepsilon^{-3}c_{ijmn}(x_1,x_2,\mathbf{x}/\varepsilon)+\varepsilon^{-2}b_{ijmn}^{(-2)}(x_1,x_2,\mathbf{x}/\varepsilon)\,, \qquad (2.3.1)$$

where

$$b_{ijmn}^{(-2)}(x_1,x_2,\mathbf{x}/\varepsilon)=\sigma_{jn}^{*(-2)}(x_1,x_2,\mathbf{x}/\varepsilon)\delta_{im} \qquad (2.3.2)$$

and

$$\langle\sigma_{\alpha\beta}^{*(-2)}(x_1,x_2,\mathbf{y})\rangle=0\ \ (\alpha,\beta=1,2)\,. \qquad (2.3.3)$$

For $h_{ijmn}(x_1,x_2,\mathbf{y})$ determined by (2.3.1), the relationship between additional stresses σ_{ij} and displacements $\mathbf{u}^{\varepsilon}$ takes the form,

$$\sigma_{ij} = [\varepsilon^{-3} c_{ijmn}^{(-3)}(\mathbf{y}) + \varepsilon^{-1} b_{ijmn}^{(-1)}(x_1, x_2, \mathbf{x}/\varepsilon)] \frac{\partial u_m^{\varepsilon}}{\partial x_n}. \tag{2.3.4}$$

Formula (2.3.4) can be considered a local governing equation.

We will comment briefly on the last term in (2.3.4) (see also comments in Sect. 2.2). The term,

$$\varepsilon^{-2} b_{ijmn}^{(-2)}(x_1, x_2, \mathbf{x}/\varepsilon) = \varepsilon^{-2} \sigma_{jn}^{*(-2)}(x_1, x_2, \mathbf{x}/\varepsilon)\delta_{im},$$

corresponds to the moments, which are independent of ε. In fact, the moments are computed by integrating the stresses multiplied by y_3 over the thickness of the plate. An integral of this kind has a characteristic value equal to ε^2.

Furthermore in this section, b_{ijmn} means $b_{ijmn}^{(-2)}$, and σ_{ij}^{*} means $\sigma_{ij}^{*(-2)}$.

Asymptotic expansions and equations of equilibrium

As above, we use the asymptotic expansion (2.2.7) and (2.2.8). As noted, the analysis of problem (2.2.1), (2.2.2), and (2.3.4) breaks down into two stages. The first entails obtaining the equations of equilibrium for the plate considered a 2-D structure. This stage is not involved with local governing equations [in the case under consideration, with (2.3.4)], and it is the same stage for any governing equations. Thus, 2-D equations of equilibrium are the same stage for the case considered in Sect. 2.2 and for the case under consideration.

The governing equations

The second stage of analysis of the problem consists of obtaining the governing equations for the plate as a 2-D structure and excluding unknown quantities from the equilibrium equations. In contrast to the first stage, this stage does involve local governing equations, and it will be investigated below.

Substituting (2.2.7) and (2.2.8) in the local governing equation (2.3.4), with allowance for (2.2.9), we obtain

$$\sum_{p=-3}^{\infty} \varepsilon^p \sigma_{ij}^{(p)} \tag{2.3.5}$$

$$= \sum_{k=0}^{\infty} \varepsilon^k [\varepsilon^{-3} c_{ijmn}(\mathbf{y}) + \varepsilon^{-2} b_{ijmn}(x_1, x_2, \mathbf{y})](u_{m,\beta x}^{(k)} \delta_{m\beta} + \varepsilon^{-1} u_{m,ny}^{(k)}).$$

Equating the terms with identical power of ε in (2.3.5), we obtain an infinite sequence of problems; the first three have the following form:

$$\sigma_{ij}^{(-3)} = c_{ijm\beta}(\mathbf{y}) u_{m,\beta x}^{(0)} + c_{ijmn}(\mathbf{y}) u_{m,ny}^{(1)}, \tag{2.3.6}$$

$$\sigma_{ij}^{(-2)} = c_{ijm\beta}(\mathbf{y}) u_{m,\beta x}^{(1)} + c_{ijmn}(\mathbf{y}) u_{m,ny}^{(2)} \tag{2.3.7}$$

$$+ b_{ijm\beta}(x_1, x_2, \mathbf{y}) u_{m,\beta x}^{(0)} + b_{ijmn}(x_1, x_2, \mathbf{y}) u_{m,ny}^{(1)},$$

$$\sigma_{ij}^{(-1)} = c_{ijm\beta}(\mathbf{y}) u_{m,\beta x}^{(2)} + c_{ijmn}(\mathbf{y}) u_{m,ny}^{(3)} \tag{2.3.8}$$

$$+ b_{ijm\beta}(x_1, x_2, \mathbf{y}) u_{m,\beta x}^{(1)} + b_{ijmn}(x_1, x_2, \mathbf{y}) u_{m,ny}^{(2)}.$$

Here, we take into account that $\mathbf{u}^{(0)}$ depends only on variables x_1, x_2.

As assumed above,

$$\mathbf{u}^{(k)}(\mathbf{y}) \text{ is periodic in } y_1, y_2 \text{ with periodicity cell } T, \quad k = 1, 2, \ldots. \tag{2.3.9}$$

Let us consider problem (2.2.24, p=–3), (2.3.6), and (2.3.9, k=1). Taking into account that $\mathbf{u}^{(0)}(x_1, x_2)$ depends only on variables x_1, x_2, we can write this problem in the form,

$$\begin{cases} [c_{ijmn}(\mathbf{y}) u_{m,ny}^{(1)} + c_{ijm\alpha}(\mathbf{y}) u_{m,\alpha x}^{(0)}(x_1, x_2)]_{,jy} = 0 & \text{in } Y, \\ [c_{ijmn}(\mathbf{y}) u_{m,ny}^{(1)} + c_{ijm\alpha}(\mathbf{y}) u_{m,\alpha x}^{(0)}(x_1, x_2)] n_j = 0 & \text{on S}, \\ \mathbf{u}^{(1)}(\mathbf{y}) \text{ is periodic in } y_1, y_2 \text{ with periodicity cell } T. \end{cases} \tag{2.3.10}$$

The solution of problem (2.3.10) can be found in the form [compare with (2.2.33)],

$$\mathbf{u}^{(1)} = \mathbf{X}^{0\alpha\beta}(\mathbf{y}) u_{\alpha,\beta x}^{(0)}(x_1, x_2) - y_3 \mathbf{e}_\alpha u_{3,\beta x}^{(0)}(x_1, x_2) + \mathbf{V}(x_1, x_2). \tag{2.3.11}$$

Here, $\mathbf{V}(x_1, x_2)$ is an arbitrary function of the argument x_1, x_2 (it will be determined below), and $\mathbf{X}^{0\alpha\beta}(\mathbf{y})$ is the solution of cellular problem (2.2.31). Substituting of (2.3.11) in (2.3.6) gives the following equation:

$$\sigma_{ij}^{(-3)} = c_{ij\alpha\beta}(\mathbf{y})u_{\alpha,\beta x}^{(0)}(x_1,x_2) + c_{ijmn}(\mathbf{y})\mathrm{X}_{m,ny}^{0\alpha\beta}(\mathbf{y})u_{\alpha,\beta x}^{(0)}(x_1,x_2). \qquad (2.3.12)$$

Averaging (2.3.12) over periodicity cell Y with $ij = \gamma\delta$, we obtain

$$\langle\sigma_{\gamma\delta}^{(-3)}\rangle = A_{\gamma\delta\alpha\beta}^{0}u_{\alpha,\beta x}^{(0)}, \qquad (2.3.13)$$

where $A_{\gamma\delta\alpha\beta}^{0}$ is given by (2.2.47).

In accordance with the homogenized equilibrium equation (2.2.17), which is the same for the cases considered in Sect. 2.2 and here,

$$(A_{\gamma\delta\alpha\beta}^{0}u_{\alpha,\beta x}^{(0)})_{,\delta x} = 0 \quad \text{in } D. \qquad (2.3.14)$$

The boundary condition for $u_{\alpha}^{(0)}$ ($\alpha = 1, 2$) follows from (2.2.2) and expansion (2.2.7):

$$u_{\alpha}^{(0)}(x_1,x_2) = 0 \ \text{ on } \partial D \ \ (\alpha = 1, 2). \qquad (2.3.15)$$

The solution of problem (2.3.14) and (2.3.15) is $u_{\alpha}^{(0)}(x_1,x_2) = 0 \ \ (\alpha = 1, 2)$. Then, (2.3.11) takes the form,

$$\mathbf{u}^{(1)} = -y_3\mathbf{e}_{\alpha}u_{3,\alpha x}^{(0)}(x_1,x_2) + \mathbf{V}(x_1,x_2). \qquad (2.3.16)$$

Substituting (2.3.16) in (2.3.6) and (2.3.7), we obtain

$$\sigma_{ij}^{(-3)} = 0, \qquad (2.3.17)$$

$$\sigma_{ij}^{(-2)} = c_{ijmn}(\mathbf{y})u_{m,ny}^{(2)} + c_{ij\alpha\beta}(\mathbf{y})y_3u_{3,\alpha x\beta x}^{(0)}(x_1,x_2) + c_{ij\alpha\beta}(\mathbf{y})V_{\alpha,\beta y}(x_1,x_2) \qquad (2.3.18)$$

$$+ b_{ij3\alpha}(x_1,x_2,\mathbf{y})u_{3,\alpha x}^{(0)}(x_1,x_2) - b_{ij\alpha 3}(x_1,x_2,\mathbf{y})u_{3,\alpha x}^{(0)}(x_1,x_2).$$

Let us examine problem (2.2.24, $p = -2$), (2.3.18), and (2.3.9, k=2). It can be written in the form,

$$\left\{\begin{array}{l} [c_{ijmn}(\mathbf{y})u^{(2)}_{m,ny} + c_{ij\alpha\beta}(\mathbf{y})y_3u^{(0)}_{3,\alpha x\beta x}(x_1,x_2) + c_{ij\alpha\beta}(\mathbf{y})V_{\alpha,\beta y}(x_1,x_2) \\ \\ + b_{ij3\alpha}(x_1,x_2,\mathbf{y})u^{(0)}_{3,\alpha x}(x_1,x_2) - b_{ij\alpha 3}(x_1,x_2,\mathbf{y})u^{(0)}_{3,\alpha x}(x_1,x_2)]_{,jy} = 0 \quad \text{in } Y, \\ \\ [c_{ijmn}(\mathbf{y})u^{(2)}_{m,ny} + c_{ij\alpha\beta}(\mathbf{y})y_3u^{(0)}_{3,\alpha x\beta x}(x_1,x_2) + c_{ij\alpha\beta}(\mathbf{y})V_{\alpha,\beta y}(x_1,x_2) \\ \\ + b_{ij3\alpha}(x_1,x_2,\mathbf{y})u^{(0)}_{3,\alpha x}(x_1,x_2) - b_{ij\alpha 3}(x_1,x_2,\mathbf{y})u^{(0)}_{3,\alpha x}(x_1,x_2)]n_j = 0 \quad \text{on } S, \\ \\ \mathbf{u}^{(2)}(\mathbf{y}) \text{ is periodic in } y_1,y_2 \text{ with perodicity cell } T. \end{array}\right. \tag{2.3.19}$$

Proposition 2.7. *Under the conditions of Proposition 2.5,*

$$b_{ij\alpha 3,jy} = 0 \quad \text{in } Y \text{ and } b_{ij\alpha 3}n_j = 0 \quad \text{on } S, \tag{2.3.20}$$

$$b_{ij3\alpha,jy} = 0 \quad \text{in } Y \text{ and } b_{ij3\alpha}n_j = 0 \quad \text{on } S.$$

In fact, $b_{ij\alpha 3,jy} = \delta_{i\alpha}\sigma^*_{j3,jy} = 0$ in Y and $b_{ij\alpha 3}n_j = \delta_{i\alpha}\sigma^*_{j3}n_j = 0$ on S by virtue of the symmetry of the initial stresses σ^*_{ij} with respect to indexes i and j and (2.2.57).

By virtue of (2.3.20), we can omit the terms containing $b_{ij3\alpha}$ and $b_{ij\alpha 3}$ in problem (2.3.19). Then, we arrive at problem (2.2.41), whose solution can be expressed through the functions $\mathbf{X}^{\nu\alpha\beta}(\mathbf{y})$ $(\nu = 0,\ 1)$ as follows [see (2.2.43)]:

$$\mathbf{u}^{(2)} = \mathbf{X}^{1\alpha\beta}(\mathbf{y})u^{(0)}_{3,\alpha x\beta x}(x_1,x_2) - y_3V_{3,\beta x}(x_1,x_2)\mathbf{e}_\beta \tag{2.3.21}$$

$$+ \mathbf{X}^{0\alpha\beta}(\mathbf{y})V_{\alpha,\beta x}(x_1,x_2).$$

Substituting (2.3.21) in (2.3.7),

$$\sigma^{(-2)}_{ij} = [c_{ijmn}(\mathbf{y})X^{0\alpha\beta}_{m,ny}(\mathbf{y}) + c_{ij\alpha\beta}(\mathbf{y})]V_{\alpha,\beta x}(x_1,x_2) \tag{2.3.22}$$

$$+ [c_{ijmn}(\mathbf{y})X^{1\alpha\beta}_{m,ny}(\mathbf{y}) - y_3c_{ij\alpha\beta}(\mathbf{y})]u^{(0)}_{3,\alpha x\beta x}(x_1,x_2)$$

$$+ b_{ij3\alpha}(x_1, x_2, \mathbf{y}) u^{(0)}_{3,\alpha x}(x_1, x_2) + b_{ij\alpha 3}(x_1, x_2, \mathbf{y}) u^{(0)}_{3,\alpha x}(x_1, x_2) .$$

Governing equations for in-plane forces and moments

Averaging (2.3.22) with $ij = \gamma\delta$ $(\gamma, \delta = 1,2)$ over periodicity cell Y, we obtain

$$N^{(-2)}_{\gamma\delta} = A^0_{\gamma\delta\alpha\beta} V_{\alpha,\beta x} + A^1_{\gamma\delta\alpha\beta} u^{(0)}_{3,\alpha x\beta x} + B^0_{\gamma\delta\alpha} u^{(0)}_{3,\alpha x} . \tag{2.3.23}$$

Multiplying (2.3.22) with $ij = \gamma\delta$ $(\gamma, \delta = 1, 2)$ by $-y_3$ and averaging over periodicity cell Y, we obtain the following equation:

$$M^{(-2)}_{\gamma\delta} = A^1_{\gamma\delta\alpha\beta} V_{\alpha,\beta x} + A^2_{\gamma\delta\alpha\beta} u^{(0)}_{3,\alpha x\beta x} + B^2_{\gamma\delta\alpha} u^{(0)}_{3,\alpha x} . \tag{2.3.24}$$

The coefficients in (2.3.23) and (2.3.24) are

$$A^{\nu+\mu}_{\gamma\delta\alpha\beta} = \langle [(-1)^\nu c_{\gamma\delta\alpha\beta}(\mathbf{y}) y_3^\nu + c_{\gamma\delta mn}(\mathbf{y}) X^{\nu\alpha\beta}_{m,ny}(\mathbf{y})](-1)^\mu y_3^\mu \rangle , \tag{2.3.25}$$

$$B^\mu_{\gamma\delta\alpha} = \langle [b_{\gamma\delta 3\alpha}(x_1, x_2, \mathbf{y}) - b_{\gamma\delta\alpha 3}(x_1, x_2, \mathbf{y})](-1)^\mu y_3^\mu \rangle$$

$$(\nu, \mu = 0, 1;\ \alpha, \beta, \gamma, \delta = 1, 2) .$$

Proposition 2.8. *Under the conditions of Proposition 2.5,* $B^\mu_{\gamma\delta\alpha} = 0$.

It follows from Proposition 2.5 and the definition of the coefficients $B^\mu_{\gamma\delta\alpha}$. The function $Z(\mathbf{y}) = \dfrac{y_3^{\mu+1}}{\mu+1}$ satisfies the conditions of Proposition 2.5. Then,

$$\langle b_{\gamma\delta\alpha n}(x_1, x_2, \mathbf{y}) Z(\mathbf{y})_{,ny} \rangle = \langle b_{\gamma\delta\alpha 3}(x_1, x_2, \mathbf{y}) y_3^\mu \rangle = 0 .$$

As follows from (2.2.28)–(2.2.30) and Proposition 2.8, in the case under consideration, the initial stresses do not affect the homogenized governing equations relating the resultant in-plane forces and moments to the in-plane strains and curvatures.

Shear forces

In order to exclude the shear forces $N^{(-1)}_{3\beta}$ and $N^{(-1)}_{\beta 3}$ from the equilibrium equations (2.2.18) and (2.2.22), we need some additional information on them.

Let us insert $\mathbf{u}^{(1)}$ into (2.3.8) in accordance with (2.3.16). Then, we obtain

$$\sigma_{ij}^{(-1)} = c_{ijm\beta}(\mathbf{y})u_{m,\beta x}^{(2)} + c_{ijmn}(\mathbf{y})u_{m,ny}^{(3)} \tag{2.3.26}$$

$$- b_{ij\beta 3}(x_1, x_2, \mathbf{y})u_{3,\beta x}^{(0)}(x_1, x_2) + b_{ij\beta 3}(x_1, x_2, \mathbf{y})u_{m,ny}^{(2)} .$$

The first two terms on the right-hand side of (2.3.26) are symmetrical with respect to i and j by virtue of the symmetry of the elastic constants (see Sect. 1.2). Then the relation below holds for the asymmetrical part of tensor $\sigma_{ij}^{(-1)}$:

$$\sigma_{ij}^{(-1)} - \sigma_{ji}^{(-1)} = [-b_{ij3\beta}(x_1, x_2, \mathbf{y}) + b_{ji3\beta}(x_1, x_2, \mathbf{y})]y_3 u_{3,\beta x}^{(0)}(x_1, x_2) \tag{2.3.27}$$

$$+ [b_{ijmn}(x_1, x_2, \mathbf{y}) + b_{jimn}(x_1, x_2, \mathbf{y})]u_{m,nx}^{(2)} .$$

Averaging equality (2.3.27) over periodicity cell Y, we obtain

$$N_{ij}^{(-1)} - N_{ji}^{(-1)} = \langle[-b_{ij3\beta}(x_1, x_2, \mathbf{y}) + b_{ji3\beta}(x_1, x_2, \mathbf{y})]y_3\rangle u_{3,\beta x}^{(0)}(x_1, x_2) \tag{2.3.28}$$

$$+ \langle[b_{ijmn}(x_1, x_2, \mathbf{y}) + b_{jimn}(x_1, x_2, \mathbf{y})]u_{m,nx}^{(2)}\rangle .$$

Proposition 2.9. *Under the conditions of Proposition 2.5, the following equality holds*:

$$\langle b_{ijmn}(x_1, x_2, \mathbf{y})u_{m,nx}^{(2)}\rangle = 0 . \tag{2.3.29}$$

Equality (2.3.29) follows from Proposition 2.5 if we set $Z = u_m^{(2)}$.

For i=3 and $j = \gamma$, we obtain from (2.3.28), with allowance for (2.3.2),

$$N_{3\gamma}^{(-1)} - N_{\gamma 3}^{(-1)} = K_\gamma , \tag{2.3.30}$$

where $K_\gamma(x_1, x_2)$ is defined by the formula

$$K_\gamma = (\langle b_{\gamma 33\beta}(x_1, x_2, \mathbf{y})y_3\rangle - \langle b_{3\gamma 3\beta}(x_1, x_2, \mathbf{y})y_3\rangle)u_{3,\beta x}^{(0)}(x_1, x_2) \tag{2.3.31}$$

$$= \langle \sigma_{\gamma\beta}^{*}(x_1, x_2, \mathbf{y})y_3\rangle u_{3,\beta x}^{(0)}(x_1, x_2) .$$

Proposition 2.10. *Under the conditions of Proposition 2.5, the following equality holds*:

$$\langle \overset{*}{\sigma}_{3\beta}(x_1, x_2, \mathbf{y}) y_3 \rangle = 0 \,. \tag{2.3.32}$$

Equality (2.3.32) follows from Proposition 2.5 if we set $Z(\mathbf{y}) = y_3^2$.

In accordance with (2.3.30), (2.3.31), and Proposition 2.10, in the case under consideration, $N_{3\gamma}^{(-1)} - N_{\gamma 3}^{(-1)} = 0$ (i.e. the shear forces are symmetrical in the indexes in spite of the fact that the corresponding local stresses are not symmetrical in the indexes). By using this fact, we can apply the classical procedure of excluding the shear forces, which leads to the classical model of a plate with no initial stresses (in spite of the fact that initial stresses exist).

The result obtained in this section shows that the initial bending/torsional moments cannot influence the deformation of a plate, in particular, they cannot cause buckling of a plate.

2.4 2-D Boundary Conditions Derived from the 3-D Boundary Problem

2-D models above were derived for a plate fastened along its boundary. In this section, we consider other types of boundary conditions of practical value and derive the corresponding boundary conditions for 2-D model of plate.

The free edge

Let a plate be fastened along the surface S_u and at the surface S_f let normal stresses $\mathbf{G}^\gamma(\mathbf{x})$ be applied to the plate (see Fig. 2.3). Denote $S_\sigma = S_\varepsilon \cup S_f$, where S_ε is the lateral surface of the plate, considered as a 3-D body.

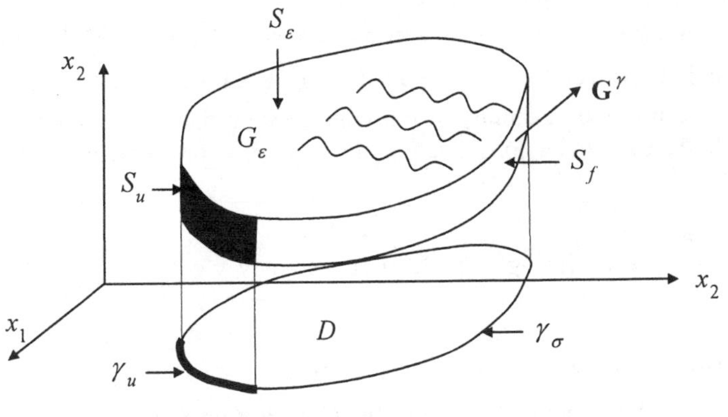

Fig. 2.3. The free edge of a plate

The corresponding sets of admissible displacement for a 3-D body and a 2-D model are the following:

$$V_\varepsilon = \{\mathbf{v} \in \{H^1(G_\varepsilon)\}^3 \ : \ \mathbf{v}(\mathbf{x}) = 0 \quad \text{on } S_u\}, \tag{2.4.1}$$

$$U_f = \{\mathbf{w}(x_1, x_2) \in \{H^1(D)\}^3 \ : \ \mathbf{w}(x_1, x_2) = 0 \quad \text{on } \gamma_u\}, \tag{2.4.2}$$

where $\gamma_\sigma \subset \partial D$ means the projection of the surface S_f to the Ox_1x_2 plane and $\gamma_u \subset \partial D$ means the projection of the surface S_u to the Ox_1x_2 plane. It is clear that $U_f \subset V_\varepsilon$.

Denote

$$\langle \circ \rangle_{S_f} = \int_{S_f} \circ \, d\mathbf{x} = \varepsilon^{-2} \int_{\varepsilon^{-1}S_f} \circ \, d\mathbf{y}.$$

the average value over the edge surface S_f.

We consider the normal stresses $\mathbf{G}^\gamma(\mathbf{x})$ applied to the surface S_f of the form $\mathbf{G}^\gamma(\mathbf{x}) = \varepsilon^{-1}\mathbf{g}^\gamma(\mathbf{x}) + \varepsilon^{-2}\mathbf{h}^\gamma(\mathbf{x})$, where $\langle \mathbf{h}^\gamma \rangle_{S_f} = 0$ and $\langle \mathbf{g}^\gamma y_3 \rangle_{Sf} \neq 0$. The stresses $\varepsilon^{-1}\mathbf{g}^\gamma$ produce the resultant boundary force and do not produce the resultant boundary moments. The thickness of the boundary is of the order of ε, then $\langle \varepsilon^{-1}\mathbf{g}^\gamma \rangle_{S_f}$ do not depend on ε. The stresses $\varepsilon^{-2}\mathbf{h}^\gamma$ produce the resultant boundary moments $\langle x_3 \varepsilon^{-2}\mathbf{h}^\gamma \rangle_{S_f} = \langle y_3 \varepsilon^{-1}\mathbf{h}^\gamma \rangle_{Sf}$, which do not depend on ε and do not produce resultant boundary forces.

In order to make the formulas shorter, we will consider the case when the mass force $\mathbf{f}^\varepsilon$ and the surface forces $\mathbf{g}^\varepsilon$ applied to the lateral surface are equal to zero. It does not restrict the generality of our consideration due to the linearity of the problem. In the case under consideration, the weak form of problem (2.2.10) is the following:

$$\int_{G_\varepsilon} \sigma_{ij} \frac{\partial v_i}{\partial x_j} d\mathbf{x} = \int_{S_f} (\varepsilon^{-1}\mathbf{g}^\gamma + \varepsilon^{-2}\mathbf{h}^\gamma)\mathbf{v} d\mathbf{x} \quad \text{for any } \mathbf{v} \in V_\varepsilon. \tag{2.4.3}$$

Here, the stresses σ_{ij} are determined by (2.2.5).

Substituting (2.2.8) in (2.4.3), we obtain, with allowance for (2.2.9),

$$\sum_{p=-2}^{\infty} \int_{G_\varepsilon} \varepsilon^p \sigma_{ij}^{(p)} (v_{i,\alpha x} \delta_{j\alpha} + \varepsilon^{-1} v_{i,jy}) d\mathbf{x} \tag{2.4.4}$$

$$= \int_{S_f} (\varepsilon^{-1} \mathbf{g}^\gamma + \varepsilon^{-2} \mathbf{h}^\gamma) \mathbf{v} d\mathbf{x} \quad \text{for any } \mathbf{v} \in V_\varepsilon .$$

We begin the summation in (2.4.4) with $p = -2$ because $\sigma_{ij}^{-3} = 0$ in accordance with (2.2.39) and (2.3.17).

We take the trial function in the form, $\mathbf{v} = \mathbf{w}(x_1, x_2) \in U_f$. Then, (2.4.4) takes the following form:

$$\sum_{p=-2}^{\infty} \int_{G_\varepsilon} \varepsilon^p \sigma_{i\alpha}^{(p)} w_{i,\alpha x} d\mathbf{x} = \int_{S_f} (\varepsilon^{-1} \mathbf{g}^\gamma + \varepsilon^{-2} \mathbf{h}^\gamma) \mathbf{w} d\mathbf{x} \quad \text{for any } \mathbf{w} \in U .$$

Taking into account Proposition 2.3 and the equality $\langle \mathbf{h}^\gamma \rangle_{S_f} = 0$, we obtain the following equality, as $\varepsilon \to 0$:

$$\sum_{p=-2}^{\infty} \int_{D} \varepsilon^{p+1} N_{i\alpha}^{(p)} w_{i,\alpha x} dx_1 dx_2 = \int_{\gamma_\sigma} \langle \mathbf{g}^\gamma \rangle_{S_f} \mathbf{w} dx_1 dx_2 \quad \text{for any } \mathbf{w} \in U_f , \tag{2.4.5}$$

where $N_{ij}^{(p)}(x_1, x_2) = \langle \sigma_{ij}^{(p)}(x_1, x_2, \mathbf{y}) \rangle$ are the resultant forces.

Integrating (2.4.5) by parts, we obtain

$$\sum_{p=-2}^{\infty} - \int_{D} \varepsilon^{p+1} N_{i\alpha}^{(p)}{}_{,\alpha x} w_i dx_1 dx_2 + \int_{\gamma_\sigma} \varepsilon^{p+1} N_{i\alpha}^{(p)} w_i n_\alpha dx_1 dx_2 \tag{2.4.6}$$

$$= \int_{\gamma_\sigma} \langle \mathbf{g}^\gamma \rangle_{S_f} \mathbf{w} dx_1 dx_2 \quad \text{for any } \mathbf{w} \in U_f .$$

Integrating (2.4.6) by parts, we obtain the equilibrium equations (2.2.16), (2.2.17) and the following boundary conditions:

$$N_{i\beta}^{(-2)} n_\beta = 0 \quad \text{on } \gamma_\sigma , \tag{2.4.7}$$

$$N_{i\beta}^{(-1)} n_\beta = \langle g_i^\gamma \rangle_{S_f} \quad \text{on } \gamma_\sigma \tag{2.4.8}$$

$(i, j = 1, 2, 3$ and $\alpha, \beta = 2, 3)$.

In the case under consideration, the boundary conditions (2.4.7) are satisfied identically.

Now take the trial function in the form, $\mathbf{v} = \mathbf{w}(x_1, x_2)y_3$. It is obvious that $\mathbf{w}(x_1, x_2)y_3 \in V_\varepsilon$. Then, (2.4.1) takes the following form:

$$\sum_{p=-2}^{\infty} \int_{G_\varepsilon} \varepsilon^p \sigma_{ij}^{(p)} (w_{i,\alpha x} \delta_{j\alpha} y_3 + \varepsilon^{-1} w_i \delta_{j3}) \mathrm{d}\mathbf{x}$$

$$= \int_{S_f} (\varepsilon^{-1} \mathbf{g}^\gamma + \varepsilon^{-2} \mathbf{h}^\gamma) \mathbf{w} y_3 \mathrm{d}\mathbf{x} \quad \text{for any } \mathbf{w} \in U_f .$$

With regard to Proposition 2.3 and equality $\langle \mathbf{g}^\gamma y_3 \rangle_{S_f}$ =0, we obtain from this equality

$$\sum_{p=-2}^{\infty} \int_D \varepsilon^{p+1} (-M_{i\alpha}^{(p)} w_{i,\alpha x} + \varepsilon^{-1} N_{i3}^{(p)} w_i) \mathrm{d}x_1 \mathrm{d}x_2 \tag{2.4.9}$$

$$= \int_{\gamma_\sigma} \varepsilon^{-1} \langle \mathbf{h}^\gamma y_3 \rangle_{S_f} \mathbf{w} \mathrm{d}x_1 \mathrm{d}x_2 \quad \text{for any } \mathbf{w} \in U_f ,$$

where the resultant moments $M_{ij}^{(p)}(x_1, x_2) = -\langle \sigma_{ij}^{(p)}(x_1, x_2, \mathbf{y}) y_3 \rangle$.

Integrating (2.4.9) by parts,

$$\sum_{p=-3}^{\infty} \int_D \varepsilon^{p+1} (-M_{i\alpha,\alpha x}^{(p)} + \varepsilon^{-1} N_{i3}^{(p)}) w_i \mathrm{d}x_1 \mathrm{d}x_2 \tag{2.4.10}$$

$$+ \int_\gamma - \varepsilon^{p+1} M_{i\alpha}^{(p)} n_\alpha w_i \mathrm{d}x_1 \mathrm{d}x_2 + \int_{\gamma_\sigma} \langle \mathbf{h}^\gamma y_3 \rangle_{S_f} \mathbf{w} \mathrm{d}x_1 \mathrm{d}x_2 = 0$$

$$\text{for any } \mathbf{w} \in U_f .$$

From (2.4.10), we obtain the equilibrium equations (2.2.21), (2.2.22), and the following 2-D boundary conditions:

$$M_{\alpha\beta}^{(-2)} n_\beta = \langle h_\alpha^\gamma y_3 \rangle_{S_f} \quad \text{on } \gamma_\sigma \tag{2.4.11}$$

(i,j=1, 2, 3 and $\alpha, \beta = 1, 2$).

Transformation of the boundary conditions

The boundary conditions (2.4.7), (2.4.8), (2.4.11) are the boundary conditions in Poisson's form. We demonstrate that they are equivalent to the boundary conditions in Kirchhoff's form.

We write the equations involving the resultant moments and shear forces from the 2-D model of a plate derived above. They are the equation of equilibrium (2.2.49), (2.2.50), and the boundary conditions (2.4.8), (2.4.11):

$$N^{(-1)}_{3\alpha\beta,\beta x} = p \quad \text{in } D, \tag{2.4.12}$$

$$M^{(-2)}_{\alpha\beta,\beta x} - N^{(-1)}_{\alpha 3} = q_\alpha \quad \text{in } D, \tag{2.4.13}$$

$$N^{(-1)}_{3\alpha} n_\alpha = n^* \quad \text{on } \gamma_\sigma, \tag{2.4.14}$$

$$M^{(-2)}_{\alpha\beta} n_\beta = m^*_\alpha \quad \text{on } \gamma_\sigma. \tag{2.4.15}$$

Here, $p = -(\langle f_3 \rangle + \langle g_3 \rangle_S)$ is the resultant normal force applied to the plate, and $q_\alpha = -(\langle f_\alpha y_3 \rangle + \langle g_\alpha y_3 \rangle_S)$ is the resultant moment of the mass and surface forces. In boundary conditions (2.4.14), (2.4.15), $n^* = \langle g^\gamma_3 \rangle_{S_f}$, $m^*_\alpha = \langle h^\gamma_\alpha y_3 \rangle_{S_f}$ are the resultant normal force and moments applied to the plate at the edge γ_σ.

If the plate has initial stresses, then, $N^{(-1)}_{3\alpha} \neq N^{(-1)}_{\alpha 3}$. Using the quantities $K_\alpha = N^{(-1)}_{\alpha 3} - N^{(-1)}_{3\alpha}$ introduced by (2.2.52), we can transform (2.4.12)–(2.4.15) into the following form:

$$(M^{(-2)}_{\alpha\beta,\beta x} + K_\alpha)_{,\alpha x} = p \quad \text{in } D, \tag{2.4.16}$$

$$M^{(-2)}_{\alpha\beta,\beta x} - N^{(-1)}_{3\alpha} + K_\alpha = q_\alpha \quad \text{in } D, \tag{2.4.17}$$

$$N^{(-1)}_{3\alpha} n_\alpha = n^* \quad \text{on } \gamma_\sigma, \tag{2.4.18}$$

$$M^{(-2)}_{\alpha\beta} n_\beta = m^*_\alpha \quad \text{on } \gamma_\sigma. \tag{2.4.19}$$

We derive from (2.4.16)–(2.4.19) the corresponding variational principle for the 2-D plate problem. Multiplying (2.4.16) by a trial function $w \in C^{\infty}(D)$ and integrating over domain D,

$$\int_D (M^{(-2)}_{\alpha\beta,\beta x} + K_{\alpha})_{,\alpha x} w \mathrm{d}x_1 \mathrm{d}x_2 = \int_D p w \mathrm{d}x_1 \mathrm{d}x_2 \text{ for any } w \in C^{\infty}(D). \quad (2.4.20)$$

Integrating (2.4.20) by parts, we obtain

$$-\int_D (M^{(-2)}_{\alpha\beta,\beta x} + K_{\alpha}) w_{,\alpha x} \mathrm{d}x_1 \mathrm{d}x_2 + \int_{\gamma_\sigma} (M^{(-2)}_{\alpha\beta,\beta x} + K_{\alpha}) w n_{\alpha} \mathrm{d}x_1 \mathrm{d}x_2$$

$$= \int_D p w \mathrm{d}x_1 \mathrm{d}x_2 \quad \text{for any } w \in C^{\infty}(D).$$

In accordance with (2.4.17), $M^{(-2)}_{\alpha\beta,\beta x} - N^{(-1)}_{3\alpha} + K_{\alpha} = q_{\alpha}$. Writing this equality on the boundary γ_σ, we obtain, with allowance for (2.4.14), the following equality:

$$(M^{(-2)}_{\alpha\beta,\beta x} + K_{\alpha}) n_{\alpha} = N^{(-1)}_{3\alpha} n_{\alpha} + q_{\alpha} n_{\alpha} = n^{*} + q_{\alpha} n_{\alpha} \quad \text{on } \gamma_\sigma. \quad (2.4.21)$$

With regard to (2.4.21), we can rewrite (2.4.20) in the form,

$$-\int_D (M^{(-2)}_{\alpha\beta,\beta x} + K_{\alpha}) w_{,\alpha x} \mathrm{d}x_1 \mathrm{d}x_2 + \int_{\gamma_\sigma} (n^{*} + q_{\alpha} n_{\alpha}) w \mathrm{d}x_1 \mathrm{d}x_2$$

$$= \int_D p w \mathrm{d}x_1 \mathrm{d}x_2 \quad \text{for any } w \in C^{\infty}(D).$$

Integrating $\int_D M^{(-2)}_{\alpha\beta,\beta x} w_{,\alpha x} \mathrm{d}x_1 \mathrm{d}x_2$ by parts, we obtain

$$\int_D (M^{(-2)}_{\alpha\beta} w_{,\alpha x\beta x} + K_{\alpha} w_{,\alpha x}) \mathrm{d}x_1 \mathrm{d}x_2 - \int_{\gamma_\sigma} M^{(-2)}_{\alpha\beta} n_{\beta} w_{,\alpha x} \mathrm{d}x_1 \mathrm{d}x_2 \quad (2.4.22)$$

$$+ \int_{\gamma_\sigma} (n^{*} + q_{\alpha} n_{\alpha}) w \mathrm{d}x_1 \mathrm{d}x_2 = \int_D p w \mathrm{d}x_1 \mathrm{d}x_2 \quad \text{for any } w \in C^{\infty}(D).$$

With regard to (2.4.19), we can write (2.4.22) in the form,

$$\int_D (M^{(-2)}_{\alpha\beta} w_{,\alpha x\beta x} + K_\alpha w_{,\alpha x})\mathrm{d}x_1\mathrm{d}x_2 - \int_{\gamma_\sigma} m^* w_{,\alpha x}\mathrm{d}x_1\mathrm{d}x_2 \tag{2.4.23}$$

$$+\int_\gamma (n^* + q_\alpha n_\alpha) w\mathrm{d}x_1\mathrm{d}x_2 = \int_D pw\mathrm{d}x_1\mathrm{d}x_2 \quad \text{for any } w \in C^\infty(D).$$

Equalities of this form are called variational principles in the book by Washizu (1982). It is known (see Washizu, 1982 for details) that from (2.4.23), one can obtain the boundary conditions in Kirchhoff's form. For $n^* = 0$, $m^* = 0$, equality (2.4.23) coincides with the variational principle for a 2-D plate with no initial stresses presented in formula (8.76) in the book by Washizu (1982).

Note that in many cases, the resultant moments of the mass and surface forces are equal to zero or may be neglected: $q_\alpha = 0$ $(\alpha = 1, 2)$. In particular, for uniform homogeneous plates the moments $q_\alpha = 0$ $(\alpha = 1, 2)$.

From the computations above, it is clear that one arrives at the boundary conditions in Poisson's form by using an arbitrary trial function $\mathbf{v}$ to write the weak form of the equilibrium equations. One arrives at the boundary conditions in Kirchhoff's form by using the trial function of the form, $\mathbf{v} = \nabla w y_3$. The last chosen is equivalent to acceptance of the Kirchhoff–Love hypothesis for a plate considered as a 3-D elastic body. However, Kirchhoff–Love hypotheses are not valid for plates of complex structure. The specifics of the approach developed above is that first we obtain the homogenized 2-D model of a plate. After that, we operate with a 2-D model and establish the equivalence of boundary conditions in Kirchhoff's form and Poisson's form for a 2-D model. Acting in this manner, we require no hypothesis about the 3-D stress-strain state.

The simply supported edge

Let a plate be fastened along the surface S_u, and at the surface S_f, it has zero normal displacements for $x_3 = 0$, and it can rotate; see Fig. 2.4. The normal stresses $\mathbf{G}^\gamma = \varepsilon^{-2}\mathbf{h}^\gamma$ are applied to the surface S_f of the plate.

The corresponding functional spaces (the sets of admissible displacement for a 3-D body and a 2-D model) are the following:

$$V_\varepsilon = \{\mathbf{v} \in \{H^1(G_\varepsilon)\}^3 : \mathbf{v}(x_1, x_2, 0) = 0 \quad \text{on } S_u\}, \tag{2.4.24}$$

$$U_f = \{\mathbf{w}(x_1, x_2) \in \{H^1(D)\}^3 : \mathbf{w}(x_1, x_2) = 0 \quad \text{on } \gamma_u\}. \tag{2.4.25}$$

It is clear that $U_f \subset V_\varepsilon$.

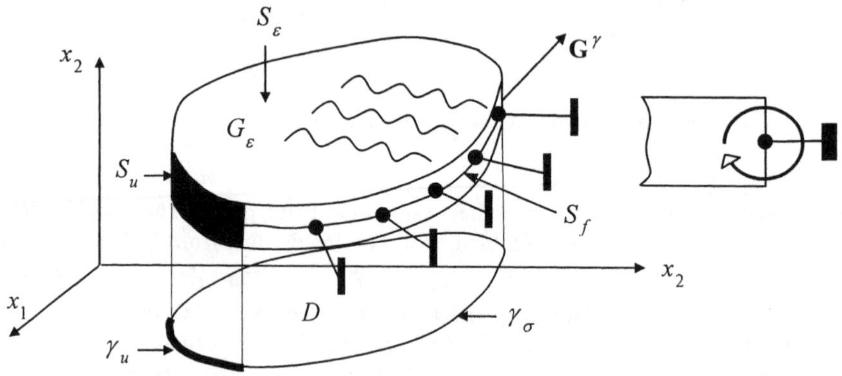

Fig. 2.4. The simply supported edge of a plate

We consider the functional space

$$U=\{\mathbf{w}(x_1,x_2)\in\{H^1(D)\}^3\ :\ \mathbf{w}(x_1,x_2)=0\ \text{ on }\partial D\}\,. \tag{2.4.26}$$

It is clear that $U\subset V_\varepsilon$. Note that the function of the form $\mathbf{w}(x_1,x_2)y_3\in V_\varepsilon$ for any $\mathbf{w}(x_1,x_2)\in U_f$. The functional spaces V_ε and U_f are determined by (2.4.1) and (2.4.2), correspondingly.

We consider the normal stresses applied to the surface S_f of the form $\varepsilon^{-2}\mathbf{h}^\gamma$. The resultant boundary moments, corresponding to these stresses, do not depend on ε.

We will consider the case when the mass force $\mathbf{f}^\varepsilon$ and the surface forces $\mathbf{g}^\varepsilon$ applied to the lateral surface are equal to zero. In this case, the weak form of (2.2.1) is the following:

$$\int\limits_{G_\varepsilon}\sigma_{ij}\frac{\partial v_i}{\partial x_j}\mathbf{dx}=\int\limits_{S_f}\varepsilon^{-2}\mathbf{h}^\gamma\mathbf{v}\mathbf{dx}\qquad\text{for any }\mathbf{v}\in W_\varepsilon\,. \tag{2.4.27}$$

We take the trial function in the form, $\mathbf{v}=\mathbf{w}(x_1,x_2)y_3$, where $\mathbf{w}(x_1,x_2)\in U$. It is obvious that $\mathbf{w}(x_1,x_2)y_3\in V_\varepsilon$. Then, (2.4.27) takes the following form:

$$\sum_{p=-2}^{\infty} \int_{G_\varepsilon} \varepsilon^p \sigma_{ij}^{(p)} (w_{i,\alpha x} \delta_{j\alpha} y_3 + \varepsilon^{-1} w_i \delta_{j3}) \mathbf{dx}$$

$$= \int_{S_f} (\varepsilon^{-1} \mathbf{g}^\gamma + \varepsilon^{-2} \mathbf{h}^\gamma) \mathbf{w} y_3 \mathbf{dx} \quad \text{for any } \mathbf{w} \in U_f .$$

With regard to Proposition 2.3, we obtain from this equation that

$$\sum_{p=-2}^{\infty} \int_{D} \varepsilon^{p+1} (M_{i\alpha}^{(p)} w_{i,\alpha x} + \varepsilon^{-1} N_{i3}^{(p)} w_i) \mathrm{d}x_1 \mathrm{d}x_2 \tag{2.4.28}$$

$$= \int_{\gamma_\sigma} \varepsilon^{-1} \langle \mathbf{h}^\gamma y_3 \rangle_{S_f} \mathbf{w} \mathrm{d}x_1 \mathrm{d}x_2 \qquad \text{for any } \mathbf{w} \in U ,$$

where $M_{\alpha\beta}^{(p)}(x_1, x_2) = -\langle \sigma_{\alpha\beta}^{(p)}(x_1, x_2, \mathbf{y}) y_3 \rangle$. Integrating (2.4.28) by parts,

$$\sum_{p=-2}^{\infty} \int_{D} \varepsilon^{p+1} (-M_{i\alpha,\alpha x}^{(p)} + \varepsilon^{-1} N_{i3}^{(p)}) w_i \mathrm{d}x_1 \mathrm{d}x_2 \tag{2.4.29}$$

$$- \int_{\gamma_\sigma} -\varepsilon^{p+1} M_{i\alpha}^{(p)} n_\alpha w_i \mathrm{d}x_1 \mathrm{d}x_2 = \varepsilon^{-1} \int_{\gamma_\sigma} \langle \mathbf{h}^\gamma y_3 \rangle_{S_f} \mathbf{w} \mathrm{d}x_1 \mathrm{d}x_2$$

$$\text{for any } \mathbf{w} \in U_f .$$

From (2.4.29), we obtain the following 2-D boundary conditions for the moments at simply supported edge:

$$M_{\alpha\beta}^{(-2)} n_\beta = \langle h_\alpha^\gamma y_3 \rangle_{S_f} \quad \text{on } \gamma_\sigma \ (i, j = 1, 2, 3 \text{ and } \alpha, \beta = 1, 2). \tag{2.4.30}$$

The complementary boundary condition is

$$u_3^{(0)}(x_1, x_2) = 0 \quad \text{on } \gamma_u . \tag{2.4.31}$$

D and 2-D "Energy Forms" for a Stressed Plate and a Stability Criterion for a Plate

The energy criterion for the stability of solids was introduced (see exposition of the energy method in Washizu, 1982) in the works by Trefftz (1930, 1933), Marguerre (1938), Prager (1947), and Hill (1958). The energy criterion is widely used in the analysis of the stability of nonlinear structures. The energy method is effective for the analysis of complex structures. In this section, the "energy form" for a 3-D thin elastic body with initial stresses will be related to the "energy form" for a homogenized 2-D plate.

"Energy forms" can be used to formulate a criterion of the stability (loss of stability) of elastic bodies. By establishing a relationship between the "energy form" for a 3-D thin elastic layer (3-D plate) and an "energy form" for a homogenized 2-D plate, we will establish the relationship between the stability of a 3-D thin elastic layer and the stability of a 2-D plate.

We define the "energy form" as

$$E(\sigma,\mathbf{u})=\frac{1}{2}\int\limits_{Q_\varepsilon}\sigma_{ij}\frac{\partial u_i^\varepsilon}{\partial x_j}\mathrm{d}\mathbf{x}\,, \qquad (2.5.1)$$

where

$$\sigma_{ij}=h_{ijmn}(x_1,x_2,\mathbf{x}/\varepsilon)\frac{\partial u_m^\varepsilon}{\partial x_n}\,,$$

$$h_{ijmn}(x_1,x_2,\mathbf{x}/\varepsilon)=\varepsilon^{-3}c_{ijmn}(\mathbf{x}/\varepsilon)$$

$$+\varepsilon^{-2}b_{ijmn}^{(-2)}(x_1,x_2,\mathbf{x}/\varepsilon)+\varepsilon^{-1}b_{ijmn}^{(-1)}(x_1,x_2,\mathbf{x}/\varepsilon)\,.$$

Asymptotic of the "energy form" in a thin layer

In accordance with the differentiating rule (2.2.9), we can rewrite (2.5.1) in the form,

$$2E(\sigma,\mathbf{u})=\int\limits_{G_\varepsilon}\sigma_{ij}(u_{i,\alpha x}^\varepsilon\delta_{j\alpha}+\varepsilon^{-1}u_{i,jy}^\varepsilon)\mathrm{d}\mathbf{x}\,. \qquad (2.5.2)$$

Substituting the asymptotic expansion for displacements (2.2.7) and the asymptotic expansion for stresses (2.2.8) in (2.5.2),

$$2E(\sigma,\mathbf{u}) = \int_{G_\varepsilon} \sum_{p=-2}^{\infty} \sum_{k=0}^{\infty} \varepsilon^{p+k} \sigma_{ij}^{(p)} (u_{i,\alpha x}^{(k)} \delta_{j\alpha} + \varepsilon^{-1} u_{i,jy}^{(k)}) \mathrm{d}\mathbf{x} \,, \tag{2.5.3}$$

Here, we take into account that $\sigma_{ij}^{(-3)} = 0$, see (2.2.39), and start the sum in p with $p = -2$. Collecting the leading terms in (2.5.3), we can write (2.5.3) in the form,

$$2E(\sigma,\mathbf{u}) = \int_{G_\varepsilon} \varepsilon^{-2} \sigma_{ij}^{(-2)} (u_{i,\alpha x}^{(0)} \delta_{j\alpha} + \varepsilon^{-1} u_{i,jy}^{(1)}) \mathrm{d}\mathbf{x} \tag{2.5.4}$$

$$+ \int_{G_\varepsilon} \varepsilon^{-1} \sigma_{ij}^{(-2)} (u_{i,\alpha x}^{(1)} \delta_{j\alpha} + \varepsilon^{-1} u_{i,jy}^{(2)}) \mathrm{d}\mathbf{x}$$

$$+ \int_{G_\varepsilon} \varepsilon^{-1} \sigma_{ij}^{(-1)} (u_{i,\alpha x}^{(0)} \delta_{j\alpha} + \varepsilon^{-1} u_{i,jy}^{(1)}) \mathrm{d}\mathbf{x} + (\ldots),$$

where $(\ldots)$ means the lowest terms (terms of the order of ε^k with $k \geq 0$).

Consider the term $u_{i,\alpha x}^{(0)} \delta_{j\alpha} + \varepsilon^{-1} u_{i,jy}^{(1)}$. Substituting $u^{(0)}(x_1, x_2) = 0$ in accordance with (2.2.37), (2.2.38) and $\mathbf{u}^{(1)}$ in accordance with (2.2.33),

$$u_{i,\alpha x}^{(0)} \delta_{j\alpha} + \varepsilon^{-1} u_{i,jy}^{(1)} = \delta_{i3} u_{3,\alpha x}^{(0)} \delta_{j\alpha} - \delta_{3j} \delta_{i\alpha} u_{3,\alpha x}^{(0)} . \tag{2.5.5}$$

Then,

$$\sigma_{ij}^{(-2)} (u_{i,\alpha x}^{(0)} \delta_{j\alpha} + \varepsilon^{-1} u_{i,jy}^{(1)}) = \sigma_{3\alpha}^{(-2)} - \sigma_{\alpha 3}^{(-2)} = 0$$

because $\sigma_{ij}^{(-2)}$ are symmetrical with respect to indexes i and j. Note that this symmetry takes place for both cases considered in Sect. 2.2 and 2.3. The first asymmetrical terms are $\sigma_{ij}^{(-1)}$. Then, (2.5.4) for the leading terms take the following form:

$$2E(\sigma,\mathbf{u}) = \int_{G_\varepsilon} \varepsilon^{-1} \sigma_{ij}^{(-2)} (u_{i,\alpha x}^{(1)} \delta_{j\alpha} + \varepsilon^{-1} u_{i,jy}^{(2)}) \mathrm{d}\mathbf{x} \tag{2.5.6}$$

$$+\int_{G_\varepsilon} \varepsilon^{-1}\sigma_{ij}^{(-1)}(u_{i,\alpha x}^{(0)}\delta_{j\alpha}+\varepsilon^{-1}u_{i,jy}^{(1)})\mathbf{dx} + (...)$$

$$=\int_{G_\varepsilon} \varepsilon^{-1}\sigma_{ij}^{(-2)}(u_{i,\alpha x}^{(1)}\delta_{j\alpha}+\varepsilon^{-1}u_{i,jy}^{(2)})\mathbf{dx}$$

$$+\int_{G_\varepsilon} \varepsilon^{-1}\sigma_{ij}^{(-1)}(\delta_{i3}u_{3,\alpha x}^{(0)}\delta_{j\alpha}-\delta_{3j}\delta_{i\alpha}u_{3,\alpha x}^{(0)})\mathbf{dx} + (...).$$

Consider the last integral in the right-hand part of (2.5.6). With regard to Proposition 2.3, it is equal to

$$\int_{G_\varepsilon}\langle\sigma_{3\alpha}^{(-1)}-\sigma_{\alpha 3}^{(-1)}\rangle u_{3,\alpha x}^{(0)}\mathrm{d}x_1\mathrm{d}x_2 = \int_D \sigma_{ij}(u_{i,\alpha x}^{\varepsilon}\delta_{j\alpha}+\varepsilon^{-1}u_{i,jy}^{\varepsilon})\mathbf{dx}.$$

In accordance with (2.2.54), the right-hand part of this equation is equal to

$$\int_D K_\alpha u_{3,\alpha x}^{(0)}\mathrm{d}x_1\mathrm{d}x_2 = \int_D N_{\alpha\beta}^{*}u_{3,\alpha x}^{(0)}u_{3,\beta x}^{(0)}\mathrm{d}x_1\mathrm{d}x_2\,. \tag{2.5.7}$$

Consider the first integral from (2.5.6):

$$\int_{G_\varepsilon}\varepsilon^{-1}\sigma_{ij}^{(-2)}(u_{i,\alpha x}^{(1)}\delta_{j\alpha}+\varepsilon^{-1}u_{i,jy}^{(2)})\mathbf{dx}\,. \tag{2.5.8}$$

In accordance with (2.2.38),

$$u_{i,\alpha x}^{(1)} = -y_3\delta_{i\beta}u_{3,\beta x\alpha x}^{(0)}(x_1,x_2)+V_{i,\beta x}(x_1,x_2), \tag{2.5.9}$$

and in accordance with (2.2.43),

$$u_{i,jy}^{(2)} = X_{i,jy}^{1\beta\alpha}(\mathbf{y})u_{3,\alpha x\beta x}^{(0)}(x_1,x_2) \tag{2.5.10}$$

$$-\delta_{j3}V_{3,\beta x}(x_1,x_2)\delta_{i\beta}+X_{i,jy}^{0\beta\alpha}(\mathbf{y})V_{\alpha,\beta x}(x_1,x_2)\,.$$

From (2.5.9) and (2.5.10), it follows that

$$u_{i,\alpha x}^{(1)}\delta_{j\alpha}+u_{i,jy}^{(2)} = -y_3\delta_{i\beta}\delta_{j\alpha}u_{3,\beta x\alpha x}^{(0)}(x_1,x_2)+\delta_{j\alpha}V_{i,\beta x}(x_1,x_2)$$

$$+X_{i,jy}^{1\beta\alpha}(\mathbf{y})u_{3,\alpha x\beta x}^{(0)}(x_1,x_2)-\delta_{j3}V_{3,\beta x}(x_1,x_2)\delta_{i\beta}$$

$$+X^{01\beta\alpha}_{i,jy}(\mathbf{y})V_{\alpha,\beta x}(x_1,x_2).$$

Grouping the terms, we obtain

$$u^{(1)}_{i,\alpha x}\delta_{j\alpha}+\varepsilon^{-1}u^{(2)}_{i,jy}=[-y_3\delta_{i\beta}\delta_{j\alpha}+X^{1\beta\alpha}_{i,jy}(\mathbf{y})]u^{(0)}_{3,\beta x\alpha x}(x_1,x_2)$$

$$+[\delta_{i\beta}\delta_{j\alpha}+X^{01\beta\alpha}_{i,jy}(\mathbf{y})]V_{i,\beta x}(x_1,x_2)$$

$$+(\delta_{i3}\delta_{j\beta}+\delta_{\beta 3}\delta_{j3})V_{3,\beta x}(x_1,x_2).$$

The quantity,

$$\sigma^{(-2)}_{ij}(\delta_{i3}\delta_{j\beta}+\delta_{\beta 3}\delta_{j3})V_{3,\beta x}(x_1,x_2)$$

$$=(\sigma^{(-2)}_{3\beta}-\sigma^{(-2)}_{\beta 3})V_{3,\beta x}(x_1,x_2)=0$$

due to the symmetry of $\sigma^{(-2)}_{ij}$. Then,

$$\sigma^{(-2)}_{ij}(u^{(1)}_{i,\alpha x}\delta_{j\alpha}+\varepsilon^{-1}u^{(2)}_{i,jy}) \tag{2.5.11}$$

$$=\sigma^{(-2)}_{ij}\{[-y_3\delta_{i\beta}\delta_{j\alpha}+X^{1\beta\alpha}_{i,jy}(\mathbf{y})]u^{(0)}_{3,\beta x\alpha x}(x_1,x_2)$$

$$+(\delta_{i\beta}\delta_{j\alpha}+X^{01\beta\alpha}_{i,jy}(\mathbf{y}))V_{i,\beta x}(x_1,x_2)\}.$$

Using (2.5.11) and substituting $\sigma^{(-2)}_{ij}$ in accordance with (2.2.44), we can write integral (2.5.8) in the form,

$$\int\limits_{G_\varepsilon}\varepsilon^{-1}\{[c_{ijmn}(\mathbf{y})X^{0\alpha\beta}_{m,ny}(\mathbf{y})+c_{ij\alpha\beta}(\mathbf{y})]V_{\alpha,\beta x}(x_1,x_2) \tag{2.5.12}$$

$$+[c_{ijmn}(\mathbf{y})X^{1\alpha\beta}_{m,ny}(\mathbf{y})-y_3c_{ij\alpha\beta}(\mathbf{y})]u^{(0)}_{3,\alpha x\beta x}(x_1,x_2)\}$$

$$\times\{[-y_3\delta_{i\beta}\delta_{j\alpha}+X^{1\beta\alpha}_{i,jy}(\mathbf{y})]u^{(0)}_{3,\beta x\alpha x}(x_1,x_2)$$

$$+[\delta_{i\beta}\delta_{j\alpha}+X^{01\beta\alpha}_{i,jy}(\mathbf{y})]V_{i,\beta x}(x_1,x_2)\}\mathrm{d}\mathbf{y}\,.$$

With allowance for Proposition 2.3 and after some transformations, we can write the equality (2.5.12) in the form

$$\int_D \langle\{[c_{ijmn}(\mathbf{y})(\delta_{m\alpha}\delta_{n\beta}+X^{0\alpha\beta}_{m,ny}(\mathbf{y})]V_{\alpha,\beta x}(x_1,x_2) \tag{2.5.13}$$

$$+[c_{ijmn}(\mathbf{y})(-\delta_{m\alpha}\delta_{n\beta}y_3+X^{1\alpha\beta}_{m,ny}(\mathbf{y})]u^{(0)}_{3,\alpha x\beta x}(x_1,x_2)\}$$

$$\times\{[-y_3\delta_{i\beta}\delta_{j\alpha}+X^{1\beta\alpha}_{i,jy}(\mathbf{y})]u^{(0)}_{3,\beta x\alpha x}(x_1,x_2)$$

$$+[\delta_{i\beta}\delta_{j\alpha}+X^{0\beta\alpha}_{i,jy}(\mathbf{y})]V_{i,\beta x}(x_1,x_2)\}\rangle\mathrm{d}x_1\mathrm{d}x_2$$

$$=\int_D[A^0_{\alpha\beta\gamma\delta}V_{\alpha,\beta x}V_{\gamma,\delta x}+(A^1_{\alpha\beta\gamma\delta}+A^1_{\gamma\delta\alpha\beta})u^{(0)}_{3,\alpha x\beta x}V_{\gamma,\delta x}$$

$$+A^2_{\alpha\beta\gamma\delta}u^{(0)}_{3,\alpha x\beta x}u^{(0)}_{3,\gamma x\delta x}]\mathrm{d}x_1\mathrm{d}x_2$$

Here, we use the equalities

$$A^0_{\alpha\beta\gamma\delta}=\langle c_{ijmn}(\mathbf{y})[\delta_{m\alpha}\delta_{n\beta}+X^{0\alpha\beta}_{m,ny}(\mathbf{y})][\delta_{i\gamma}\delta_{j\delta}+X^{0\gamma\delta}_{m,ny}(\mathbf{y})]\rangle\,,$$

$$A^1_{\alpha\beta\gamma\delta}=\langle c_{ijmn}(\mathbf{y})[-\delta_{m\alpha}\delta_{n\beta}y_3+X^{1\alpha\beta}_{m,ny}(\mathbf{y})][\delta_{i\gamma}\delta_{j\delta}+X^{0\gamma\delta}_{m,ny}(\mathbf{y})]\rangle\,,$$

$$A^2_{\alpha\beta\gamma\delta}=\langle c_{ijmn}(\mathbf{y})[-\delta_{m\alpha}\delta_{n\beta}y_3+X^{1\alpha\beta}_{m,ny}(\mathbf{y})][-\delta_{i\gamma}\delta_{j\delta}y_3+X^{1\gamma\delta}_{m,ny}(\mathbf{y})]\rangle\,,$$

which will be proved in Sect. 4.1 devoted to computation of stiffnesses of inhomogeneous plates with no initial stresses.

As a result, we find that the leading term of the "energy form" $E(\sigma,\mathbf{u})$ (2.5.1) of an inhomogeneous plate considered as a 3-D thin body is of the order $\varepsilon^0=1$, and it is equal to the sum of (2.5.7) and (2.5.13):

$$E(\sigma,\mathbf{u}) = E(u_3^{(0)}, V_1, V_2) + \frac{1}{2}\int_D N^*_{\alpha\beta} u^{(0)}_{3,\alpha x} u^{(0)}_{3,\beta x} \mathrm{d}x_1 \mathrm{d}x_2 , \tag{2.5.14}$$

where $E(u_3, V_1, V_2)$ is the energy of a homogenized 2-D plate with no initial stresses:

$$E(u_3^{(0)}, V_1, V_2) = \frac{1}{2}\int_D [A^0_{\alpha\beta\gamma\delta} V_{\alpha,\beta x} V_{\gamma,\delta x} + (A^1_{\alpha\beta\gamma\delta} + A^1_{\gamma\delta\alpha\beta}) u^{(0)}_{3,\alpha x\beta x} V_{\gamma,\delta x}$$

$$+ A^2_{\alpha\beta\gamma\delta} u^{(0)}_{3,\alpha x\beta x} u^{(0)}_{3,\gamma x\delta x}]\mathrm{d}x_1 \mathrm{d}x_2 .$$

In accordance with (2.5.14), the "energy form" $E(\sigma,\mathbf{u})$ (2.5.1) of a thin 3-D elastic layer with initial stresses is equal to the "energy form" $E(u_3^{(0)}, V_1, V_2)$ for a 2-D plate. This equality makes it possible to use the results obtained in Sect. 2.2 for the analysis of the stability of plates of complex structure. In the works by Trefftz (1930, 1933), Marguerre (1938), Prager (1947), and Hill (1958), a criterion of elastic stability was formulated in terms of the "energy form." In the case under consideration, the criterion mentioned states the following: if for given $\{N^*_{\alpha\beta}\}$, the "energy form,"

$$E(\sigma,\mathbf{u}) = E(u_3^{(0)}, V_1, V_2) + \frac{1}{2}\int_D N^*_{\alpha\beta} u^{(0)}_{3,\alpha x} u^{(0)}_{3,\beta x} \mathrm{d}x_1 \mathrm{d}x_2 \geq 0 ,$$

for any admissible displacements, then the body is stable; if the "energy form,"

$$E(\sigma,\mathbf{u}) = E(u_3^{(0)}, V_1, V_2) + \frac{1}{2}\int_D N^*_{\alpha\beta} u^{(0)}_{3,\alpha x} u^{(0)}_{3,\beta x} \mathrm{d}x_1 \mathrm{d}x_2 ,$$

takes both positive and negative values, then the body is not stable under the action of the in-plane resultant stresses $\{N^*_{\alpha\beta}\}$, and there exist nonzero displacements $u_3^{(0)}, V_1, V_2$, which satisfy the equations,

$$\delta_{V_\beta} E(u_3^{(0)}, V_1, V_2) = 0, \ (\beta = 1, 2), \tag{2.5.15}$$

$$\delta_{u_3^{(0)}} E(u_3^{(0)}, V_1, V_2) = 0,$$

where δ means the variation of functional $E(u_3^{(0)}, V_1, V_2)$ (2.5.14) with respect to the corresponding variable.

Buckling problem

Let us write (2.5.15) for the special case, when the additional forces are equal to zero: $\mathbf{f}^{\varepsilon}=0$, **g**=0; the initial stresses are proportional to a parameter λ and satisfy the conditions of Proposition 2.5. The variational equality (2.5.15) can be written in the following form:

$$(A^0_{\gamma\delta\alpha\beta}V_{\alpha,\beta x} + A^1_{\gamma\delta\alpha\beta}u^{(0)}_{3,\alpha x\beta x})_{,\delta x} = 0 \quad \text{in } D, \tag{2.5.16}$$

$$(A^1_{\gamma\delta\alpha\beta}V_{\alpha,\beta x} + A^2_{\gamma\delta\alpha\beta}u^{(0)}_{3,\alpha x\beta x})_{,\delta x} = \lambda(N^*_{\delta\beta}u^{(0)}_{3,\beta x})_{,\delta x} \quad \text{in } D, \tag{2.5.17}$$

$$V_\alpha(x_1,x_2) = 0 \;\; (\alpha = 1,2),\; u^{(0)}_3(x_1,x_2) = \frac{\partial u^{(0)}_3}{\partial x_i}(x_1,x_2)n_i = 0 \quad \text{on } \partial D. \tag{2.5.18}$$

Problem (2.5.16)–(2.5.18) is an eigenvalue problem, describing the buckling of stressed plate. If $A^1_{\gamma\delta\alpha\beta} \neq 0$, (2.5.16) and (2.5.17) are coupled. If $A^1_{\gamma\delta\alpha\beta} = 0$, $V_\alpha(x_1,x_2) = 0$ ($\alpha = 1,2$) in D, and problem (2.5.16)–(2.5.18) takes the form,

$$(A^2_{\gamma\delta\alpha\beta}u^{(0)}_{3,\alpha x\beta x})_{,\delta x} = \lambda(N^*_{\delta\beta}u^{(0)}_{3,\beta x})_{,\delta x} \quad \text{in } D, \tag{2.5.19}$$

$$u^{(0)}_3(x_1,x_2) = \frac{\partial u^{(0)}_3}{\partial x_i}(x_1,x_2)n_i = 0 \quad \text{on } \partial D. \tag{2.5.20}$$

Model (2.5.19), (2.5.20) coincides qualitatively with the classical equations.

2.6 Membrane (2-D Model)

Problems involving an inhomogeneous membrane, in which both the inhomogeneities of structure and the initial stresses play a decisive role, touch directly on problems of the above type.

We will term a membrane a thin body whose behavior is determined by in-plane initial stresses. For membranes, whose thickness is significantly less than the length of the period, 2-D equations of the theory of thin inhomogeneous plates may be taken as the initial ones.

Formulation of the problem

We will examine a 2-D body of periodic structure obtained by repeating a certain small periodicity cell εY over the Ox_1x_2 plane (Fig. 2.5). Here ε is the periodicity cell's characteristic dimension, which is assumed to be small (that is formalized in the form, $\varepsilon \to 0$).

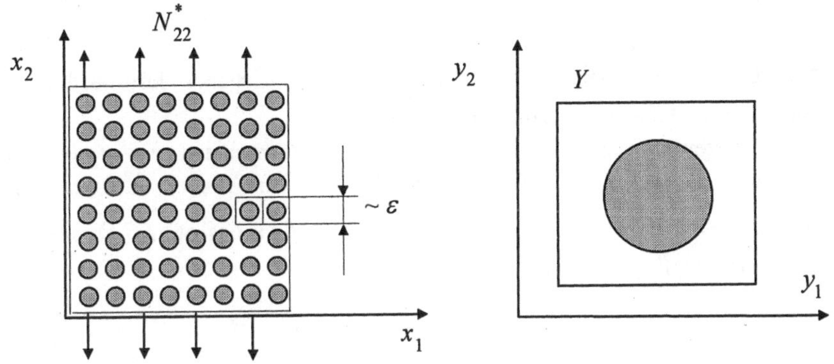

Fig. 2.5. 2-D membrane of periodic structure.

The starting point of our investigation is a 2-D formulation of a membrane problem. In accordance with this model, the equilibrium equations of the membrane can be written in the form,

$$\frac{\partial M_i}{\partial x_i} = f(\mathbf{x}, \mathbf{x}/\varepsilon) \quad \text{in } G, \tag{2.6.1}$$

where

$$M_i = \sigma^*_{ij}(\mathbf{x}, \mathbf{x}/\varepsilon)\frac{\partial w^\varepsilon}{\partial x_j}. \tag{2.6.2}$$

The boundary conditions can be written in the form,

$$w^\varepsilon(\mathbf{x})=0 \text{ on } \partial G. \tag{2.6.3}$$

Here, G is a 2-D region occupied by the membrane, the membrane is fastened over the boundary ∂G (see Fig. 2.5), $\mathbf{x} = (x_1, x_2)$, w^ε are the normal deflection, $\sigma^*_{ij}(\mathbf{x}, \mathbf{x}/\varepsilon)$ are the initial stresses, and $f(\mathbf{x}, \mathbf{x}/\varepsilon)$ is the normal force (a sum of the mass force multiplied by the thickness of the plate and the normal surface forces).

Asymptotic expansions and equations of equilibrium

We will study global deformations of the membrane as $\varepsilon \to 0$. To do this, we use the following asymptotic expansion: asymptotic expansion for deflection,

$$w^{\varepsilon} = w^{(0)}(\mathbf{x}) + \varepsilon^1 w^{(1)}(\mathbf{x},\mathbf{y}) + \ldots = \sum_{k=0}^{\infty} \varepsilon^k w^{(k)}(\mathbf{x},\mathbf{y}), \tag{2.6.4}$$

asymptotic expansion for resultant stresses,

$$M_i = \sum_{k=0}^{\infty} \varepsilon^k M_i^{(k)}(\mathbf{x},\mathbf{y}). \tag{2.6.5}$$

Note that in the expansion for deflection, $w^{(0)}(\mathbf{x})$ depends only on the "slow" variable $\mathbf{x}$ and the term according to ε is absent.

The differential operators are presented in the form of the sum of operators in $\{x_i\}$ and in $\{y_i\}$ (see Sect. 1.2). For the function $Z(\mathbf{x},\mathbf{y})$ of the arguments $\mathbf{x} = (x_1, x_2)$ and $\mathbf{y} = (y_1, y_2)$, as on the right–hand sides of (2.6.4) and (2.6.5), this representation takes the form,

$$\frac{\partial Z}{\partial x_i} = Z_{,ix} + \varepsilon^{-1} Z_{,iy} \quad (i=1,\,2). \tag{2.6.6}$$

Substituting (2.6.4) and (2.6.5) in (2.6.1), we obtain, with allowance for (2.6.6),

$$\sum_{k=0}^{\infty} \varepsilon^k (M_{i,ix}^{(k)} + \varepsilon^{-1} M_{i,iy}^{(k)}) = f(\mathbf{x},\mathbf{y}). \tag{2.6.7}$$

Averaging (2.6.7) with regard to Proposition 2.1 and equating terms of the same order of ε, we obtain an infinite sequence of homogenized equilibrium equations; the first is the following:

$$\langle M_i^{(0)} \rangle_{,ix} = \langle f \rangle(\mathbf{x}). \tag{2.6.8}$$

We use here the equality $\langle Z_{,iy} \rangle = 0$ for every Y-periodic function $Z(\mathbf{y})$.

Equating terms of the same order of ε in (2.6.7), we obtain an infinite sequence of local equilibrium equations; the first is the following:

$$M_{i,iy}^{(0)} = 0. \tag{2.6.9}$$

Governing equations

Substituting (2.6.4), (2.6.5) in the local governing equation (2.6.2) and replacing the differential equations in accordance with (2.6.6),

$$\sum_{k=0}^{\infty} \varepsilon^k M_i^{(k)} = \sum_{k=0}^{\infty} \varepsilon^k \overset{*}{\sigma}_{ij} (w_{,jx}^{(k)} + \varepsilon^{-1} w_{,jy}^{(k)}) . \tag{2.6.10}$$

Equating terms of the order of ε^0 in (2.6.10), we obtain

$$M_i^{(0)} = \overset{*}{\sigma}_{ij}(\mathbf{y}) w_{,jx}^{(1)} + \overset{*}{\sigma}_{ij}(\mathbf{y}) w_{,jy}^{(0)}(\mathbf{x}) . \tag{2.6.11}$$

Substituting (2.6.11) in (2.6.9),

$$[\overset{*}{\sigma}_{ij}(\mathbf{y}) w_{,jy}^{(1)} + \overset{*}{\sigma}_{ij}(\mathbf{y}) w_{,jx}^{(0)}(\mathbf{x})]_{,jy} = 0 . \tag{2.6.12}$$

The solution of problem (2.6.12) with periodicity conditions is

$$w^{(1)} = K^m(\mathbf{y}) w_{,mx}^{(0)}(\mathbf{x}) + V(\mathbf{x}) , \tag{2.6.13}$$

where $K^m(\mathbf{y})$ is the solution of the following cellular problem:

$$\begin{cases} [\overset{*}{\sigma}_{ij}(\mathbf{y}) K_{,jy}^m + \overset{*}{\sigma}_{im}(\mathbf{y})]_{,jy} = 0 \quad \text{in } Y, \\ K^m(\mathbf{y}) \text{ periodic in } \mathbf{y} \text{ with periodicity cell } Y. \end{cases} \tag{2.6.14}$$

Substituting (2.6.13) in (2.6.11), we obtain

$$M_i^{(0)} = [\overset{*}{\sigma}_{ij}(\mathbf{y}) K_{,jy}^m(\mathbf{y}) + \overset{*}{\sigma}_{jm}(\mathbf{y})] w_{,mx}^{(0)}(\mathbf{x}) . \tag{2.6.15}$$

The limit problem

Averaging (2.6.15) over the periodicity cell Y, with allowance for Proposition 2.1,

$$M_i^{(0)} = \Sigma_{im} w_{,mx}^{(0)}(\mathbf{x}) , \tag{2.6.16}$$

where

$$\Sigma_{im} = \langle \overset{*}{\sigma}_{ij}(\mathbf{y}) K_{,jy}^m(\mathbf{y}) + \overset{*}{\sigma}_{jm}(\mathbf{y}) \rangle . \tag{2.6.17}$$

It is seen from (2.6.17) that $\Sigma_{im} \neq \langle \overset{*}{\sigma}_{jm}(\mathbf{y}) \rangle$ in the general case.

The boundary condition for the limit problem can be obtained by substituting the expansion (2.6.4) in the original boundary conditions (2.6.3):

$$w^{(0)}(\mathbf{x}) = 0 \quad \text{on } \partial G . \tag{2.6.18}$$

The limit problem (2.14.16) and (2.6.17) can be written as

$$\langle \Sigma_{im} w^{(0)}_{,mx} \rangle_{,ix} = \langle f \rangle (\mathbf{x}) . \tag{2.6.19}$$

The results obtained (underline that it is a formal result) means that the behavior of an inhomogeneous membrane is determined not by the in-plane stresses $\langle \sigma^*_{jm}(\mathbf{y}) \rangle$, but the quantities Σ_{im}. This fact is not mentioned in the literature in mechanics. In the next section, we will consider a more rigorous model of a membrane derived directly from the 3-D elasticity model and demonstrate that the behavior of the membrane is determined the in-plane stresses $\langle \sigma^*_{jm}(\mathbf{y}) \rangle$.

Formula (2.6.17) is a result of a formal analysis of 2-D classical model written for a nonhomogeneous membrane. It shows that the application of mathematical methods must be coupled with mechanical analysis of a problem. It also shows that the classical 2-D model of a membrane is not universal. In the next section, the contradictions in the classical model of a membrane will be discussed.

2.7 Membranes (Transition from 3-D to 2-D Model)

In the previous sections, an asymptotic method of converting a 3-D elasticity problem for a thin domain into a 2-D plate problem was described. The advantage of this method is that it is applicable to both classical homogeneous and slightly inhomogeneous plates as well as to inhomogeneous plates of a periodic structure having inhomogeneities, whose dimensions are comparable with the plate thickness.

In this section, a 2-D membrane model is derived from a 3-D elasticity problem. There is a certain resemblance between the methods used in Sect. 2.2 and those used in this section, but the results obtained do not follow from the results presented in Sect. 2.2. This is due to the fact that the "mechanics" of a problem is governed mainly by the actual form of the first few terms in the asymptotic expansion. These terms depend on the order of initial stresses compared with the elastic constants.

Examples of the practical application of inhomogeneous membranes can be found in the book by Chou and Ko (1989).

Statement of the problem

We consider a domain G_ε that is obtained by periodic repetition of periodicity cell εY in the Ox_1x_2 plane; see Fig. 2.6.

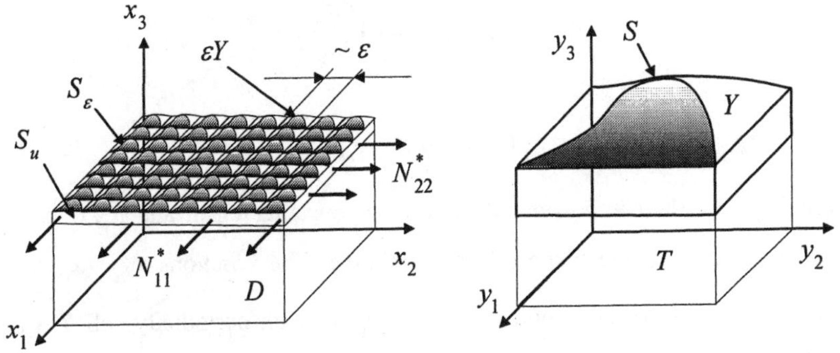

Fig. 2.6. Thin layer with initial stresses (3-D membrane) of periodic structure and its periodicity cell

The characteristic size of this periodicity cell (which is identical to the characteristic thickness of the membrane) is a small quantity $\varepsilon << 1$ that is formalized in the form $\varepsilon \to 0$. When $\varepsilon \to 0$, domain G_ε "tightens" to a 2-D domain D in the Ox_1x_2 plane.

We assume that the structure under consideration has initial stresses and these stresses, denoted by σ^*_{ij}, are determined from the solution of the 3-D problem of the theory of elasticity.

We shall assume that the elasticity constants of the membrane material c_{ijmn} and initial stresses σ^*_{ij} are of the same order ε^{-1}. Note that in classical theory, the in-plane stiffnesses of the membrane are neglected. In the more rigorous theory, we cannot neglect the stiffnesses due to the fact that it is impossible to create initial stresses σ^*_{ij} of a greater order of magnitude then the elastic constants c_{ijmn} when the plate is deformed in its plane.

The equilibrium equations for a body with initial stresses have the form (see Sect. 1.1),

$$\frac{\partial \sigma_{ij}}{\partial x_j} + \varepsilon^{-1} f_i(x_1, x_2, \mathbf{x}/\varepsilon) = 0 \quad \text{in } G_\varepsilon, \tag{2.7.1}$$

$$\sigma_{ij} n_j = g_i(x_1, x_2, \mathbf{x}/\varepsilon) \quad \text{on } S_\varepsilon,$$

$$\mathbf{u}^\varepsilon(\mathbf{x}) = 0 \quad \text{on } S_u,$$

where S_ε and S_u are the surfaces of domain G_ε (see Fig. 2.6).

The relation between current stresses σ_{ij}, displacements $\mathbf{u}^\varepsilon$, and initial stresses σ^*_{ij} may be written in the form (see Sect. 1.1),

$$\sigma_{ij} = \varepsilon^{-1} h_{ijmn}(x_1, x_2, \mathbf{x}/\varepsilon) \frac{\partial u^\varepsilon_m}{\partial x_n}, \tag{2.7.2}$$

$$h_{ijmn}(x_1, x_2, \mathbf{x}/\varepsilon) = c_{ijmn}(x_1, x_2, \mathbf{x}/\varepsilon) + \sigma^*_{jn}(x_1, x_2, \mathbf{x}/\varepsilon)\delta_{im}, \tag{2.7.3}$$

where c_{ijmn} are the components of the elastic constants tensor and x_1, x_2 are the "slow" variables in the plane of the membrane. The functions $c_{ijmn}(x_1, x_2, \mathbf{y})$, $\sigma^*_{jn}(x_1, x_2, \mathbf{y})$ are periodic with respect to y_1, y_2 with periodicity cell T (T is the projection of the periodicity cell εY on the Oy_1y_2 plane; see Fig. 2.6).

Asymptotic expansions

We introduce "fast" variables $\mathbf{y} = \mathbf{x}/\varepsilon$. The derivative of a function of the form, $Z(x_1, x_2, \mathbf{y})$, is calculated by replacing the differential operator according to the rule,

$$\frac{\partial Z}{\partial x_3} = \varepsilon^{-1} Z_{,3y}, \tag{2.7.4}$$

$$\frac{\partial Z}{\partial x_\alpha} = Z_{,\alpha x} + \varepsilon^{-1} Z_{,\alpha y} \quad (\alpha = 1, 2).$$

The Latin indexes take values 1,2,3, and the Greek indexes take values 1,2; m=–1, 0,...; n=0, 1,...; subscript $,\alpha x$ means $\partial/\partial x_\alpha$, and subscript $,3y$ means $\partial/\partial x_3$.

We shall seek a solution of problem (2.7.1)–(2.7.3) in a form, which is analogous to that used previously:

$$\mathbf{u}^\varepsilon = \mathbf{u}^{(0)}(x_1, x_2) + \sum_{k=1}^{\infty} \varepsilon^k \mathbf{u}^{(k)}(x_1, x_2, \mathbf{y}). \tag{2.7.5}$$

In accordance with (2.7.2), we start the expansion for stresses σ_{ij} with a term of the order of ε^{-1}:

$$\sigma_{ij} = \sum_{p=-1}^{\infty} \varepsilon^p \sigma_{ij}^{(p)}(x_1, x_2, \mathbf{y}) . \tag{2.7.6}$$

Substituting (2.7.5) in (2.7.2) and taking into account (2.7.4),

$$\sum_{p=-1}^{\infty} \varepsilon^p \sigma_{ij}^{(p)} = \sum_{k=0}^{\infty} \varepsilon^{k-1} h_{ijmn}(x_1, x_2, \mathbf{y}) u_{m,ny}^{(k)} + \varepsilon^k h_{ijm\alpha}(x_1, x_2, \mathbf{y}) u_{m,\alpha x}^{(k)} . \tag{2.7.7}$$

Equating the expressions accompanying ε^{-1} in (2.7.7), we obtain

$$\sigma_{ij}^{(-1)} = h_{ijmn}(x_1, x_2, \mathbf{y}) u_{m,ny}^{(1)} + h_{ijm\alpha}(x_1, x_2, \mathbf{y}) u_{m,\alpha x}^{(0)}(x_1, x_2) . \tag{2.7.8}$$

Consider the equilibrium equations (2.7.1). Substituting (2.7.6) in (2.7.1), with allowance (2.7.4) and equating the terms accompanying the same powers of ε, we obtain the terms of expansion (2.7.6) that satisfy the equations (see Sect. 2.2)

$$\sigma_{ij,jy}^{(m)} + \sigma_{i\alpha,\alpha x}^{(m-1)} = 0 , \tag{2.7.9}$$

(we assume $\sigma_{ij}^{(-2)} = 0$).

Only the case m=1 will be interesting for us. For this case, the equation (2.7.9) takes the form,

$$\sigma_{ij,jy}^{(-1)} = 0 . \tag{2.7.10}$$

Substituting (2.7.8) in (2.7.10), we arrive in the usual way at cellular problem

$$\begin{cases} [h_{ijkl}(x_1, x_2, \mathbf{y}) K_{k,ly}^{0m\alpha} + h_{ijm\alpha}(x_1, x_2, \mathbf{y})]_{,jy} = 0 & \text{in } Y , \\ [h_{ijkl}(x_1, x_2, \mathbf{y}) K_{k,ly}^{0m\alpha} + h_{ijm\alpha}(x_1, x_2, \mathbf{y})] n_j = 0 & \text{on } S, \\ \mathbf{K}^{0m\alpha}(\mathbf{y}) \text{ is periodic in } y_1, y_2 \text{ with periodicity cell } T \end{cases} \tag{2.7.11}$$

($\alpha, \beta = 1, 2$ and m=1, 2, 3).

Here, S means the lateral surface of periodicity cell Y (Fig. 2.6).

As suggested above, σ_{ij}^{*} are determined from the solution of a problem in the theory of elasticity

$$\begin{cases} \dfrac{\partial \overset{*}{\sigma}_{ij}}{\partial x_j}(\mathbf{x}, \mathbf{x}/\varepsilon) = 0 \quad \text{in } G_\varepsilon, \\ \overset{*}{\sigma}_{ij} n_j = g_i(\mathbf{x}/\varepsilon) \quad \text{on } S_\varepsilon, \\ \mathbf{u}^\varepsilon(\mathbf{x}) = 0 \quad \text{on } S_u. \end{cases}$$

Allowing for the differentiating rule (2.7.4), from these equations,

$$\begin{cases} \overset{*}{\sigma}_{jl,jy} = 0 \ \text{in } Y, \\ \overset{*}{\sigma}_{jl} n_j = 0 \ \text{on } S. \end{cases}$$

Using these equalities and the definition of h_{ijmn} in (2.7.3),

$$[h_{ijmn}(x_1, x_2, \mathbf{y}) - c_{ijmn}(\mathbf{y})]_{,jy} = \overset{*}{\sigma}_{jn,jy}(x_1, x_2, \mathbf{y})\delta_{im} = 0 \quad \text{in } Y, \tag{2.7.12}$$

$$[h_{ijmn}(x_1, x_2, \mathbf{y}) - c_{ijmn}(\mathbf{y})] n_j = \overset{*}{\sigma}_{jn}(x_1, x_2, \mathbf{y}) n_j \delta_{im} = 0 \quad \text{on } S.$$

Proposition 2.12. *The following equality holds:*

$$\mathbf{K}^{0m\alpha}(\mathbf{y}) = -y_3 \mathbf{e}_\alpha. \tag{2.7.13}$$

Proposition 2.12 is an analog of Proposition 2.11.

Taking into account Proposition 2.12, we obtain a representation of $\mathbf{u}^{(1)}$ in terms of the solution of cellular problem (2.7.11):

$$\mathbf{u}^{(1)} = \mathbf{K}^{0\beta\alpha}(\mathbf{y}) u^{(0)}_{\beta,\alpha x}(x_1, x_2) - y_3 u^{(0)}_{3,\alpha x}(x_1, x_2) + \mathbf{V}(x_1, x_2). \tag{2.7.14}$$

This representation is similar in its form to representation (2.2.33) from Sect. 2.2, but the coefficients A_{ijmn} of cellular problem (2.7.11) differ from the elasticity constants c_{ijmn} and depend on the initial stresses.

Substituting (2.7.14) in (2.7.8) gives

$$\sigma^{(-1)}_{ij} = [-h_{ij\alpha 3}(x_1, x_2, \mathbf{y}) + h_{ij3\alpha}(x_1, x_2, \mathbf{y})] u^{(0)}_{3,\alpha x}(x_1, x_2) \tag{2.7.15}$$

$$+[h_{ij\alpha\beta}(x_1,x_2,\mathbf{y})+h_{ijmn}(x_1,x_2,\mathbf{y})K^{0\beta\alpha}_{m,ny}]u^{(0)}_{\beta,\alpha x}(x_1,x_2)\,.$$

By using the definition of h_{ijmn} (2.7.3) and the symmetry of c_{ijmn}, we obtain from (2.7.15)

$$\sigma^{(-1)}_{ij}=[-\overset{*}{\sigma}_{j3}(x_1,x_2,\mathbf{y})\delta_{i\alpha}+\overset{*}{\sigma}_{j\alpha}(x_1,x_2,\mathbf{y})\delta_{i3}]u^{(0)}_{3,\alpha x}(x_1,x_2) \tag{2.7.16}$$

$$+[h_{ij\alpha\beta}(x_1,x_2,\mathbf{y})+h_{ijmn}(x_1,x_2,\mathbf{y})K^{0\beta\alpha}_{m,ny}]u^{(0)}_{\beta,\alpha x}(x_1,x_2)\,.$$

The limit 2-D model of a membrane

As above,

$$\langle\circ\rangle=(mesT)^{-1}\int_Y\circ\,d\mathbf{y}\ \text{ and }\ \langle\circ\rangle_S=(mesT)^{-1}\int_S\circ\,d\mathbf{y}$$

are average values over periodicity cell Y and over the lateral surface S of periodicity cell Y in "fast" variables $\mathbf{y}=\mathbf{x}/\varepsilon$ (Fig. 2.6).

Equilibrium equations

In the case under consideration, the equilibrium equations for in-plane stresses $N_{ij}=\langle\sigma^{(-1)}_{ij}\rangle$ are identical to those obtained in Sect. 2.2 and have the form,

$$N_{\alpha\gamma,\gamma x}+\langle f_\alpha\rangle(x_1,x_2)=0, \tag{2.7.17}$$

$$N_{3\gamma,\gamma x}+\langle f_3\rangle(x_1,x_2)+\langle g_3\rangle_S(x_1,x_2)=0\,.$$

Constitutive relations

Consider (2.7.16) for different values of the indexes. When i=3, $j=\gamma$, (2.7.16) taking (2.7.3), into account, gives

$$N_{3\gamma}=\langle\overset{*}{\sigma}_{\gamma\alpha}(x_1,x_2,\mathbf{y})\rangle u^{(0)}_{3,\alpha x}(x_1,x_2) \tag{2.7.18}$$

$$+\langle h_{3\gamma\alpha\beta}(x_1,x_2,\mathbf{y})+h_{3\gamma mn}(x_1,x_2,\mathbf{y})K^{0\beta\alpha}_{m,ny}(\mathbf{y})\rangle u^{(0)}_{\beta,\alpha x}(x_1,x_2)\,.$$

When $i=\kappa$, $j=\gamma$ ($\kappa,\gamma=1,2$), (2.7.16) gives

$$N_{\kappa\gamma} = \langle -\sigma^*_{\gamma 3}(x_1, x_2, \mathbf{y})\rangle u^{(0)}_{3,\kappa\alpha}(x_1, x_2) \tag{2.7.19}$$

$$+ \langle h_{\kappa\gamma\alpha\beta}(x_1, x_2, \mathbf{y}) + h_{\kappa\gamma mn}(x_1, x_2, \mathbf{y}) K^{0\beta\alpha}_{m,ny}(\mathbf{y})\rangle u^{(0)}_{\beta,\alpha x}(x_1, x_2) .$$

We write (2.7.18) and (2.7.19) in the form,

$$N_{3\gamma} = N^*_{\gamma\alpha} u^{(0)}_{3,\alpha x}(x_1, x_2) + R_{\gamma\beta\alpha} u^{(0)}_{\beta,\alpha x}(x_1, x_2) , \tag{2.7.20}$$

$$N_{\kappa\gamma} = C_{\kappa\gamma\beta\alpha}(\sigma) u^{(0)}_{\beta,\alpha x}(x_1, x_2) , \tag{2.7.21}$$

where

$$N^*_{\gamma\alpha} = \langle \sigma^*_{\gamma\alpha}(x_1, x_2, \mathbf{y})\rangle ,$$

$$R_{\gamma\beta\alpha} = \langle h_{3\gamma\alpha\beta}(x_1, x_2, \mathbf{y}) + h_{3\gamma mn}(x_1, x_2, \mathbf{y}) K^{0\beta\alpha}_{m,ny}(\mathbf{y})\rangle ,$$

$$C_{\kappa\gamma\alpha\beta}(\sigma) = \langle h_{\kappa\gamma\alpha\beta}(x_1, x_2, \mathbf{y}) + h_{\kappa\gamma mn}(x_1, x_2, \mathbf{y}) K^{0\beta\alpha}_{m,ny}(\mathbf{y})\rangle$$

$$(\kappa, \gamma, \alpha, \beta = 1, 2) .$$

Here, $N^*_{\gamma\alpha}$ are the resultant in-plane initial stresses in the membrane, and $C_{\kappa\gamma\alpha\beta}$ $(\kappa, \gamma, \alpha, \beta = 1,2)$ are the homogenized elastic constants of the stressed body (which is two-dimensional in the case under consideration). The constants $C_{\kappa\gamma\alpha\beta}$, generally, depend on the initial stresses in the plane of the membrane. This dependence is an analog of the dependence found by Kolpakov (1990, 1992b).

The boundary conditions

The boundary conditions in the terms of displacements can be written in the form

$$u^{(0)}_\alpha(x_1, x_2) = 0 \quad \text{on } \partial D \ (\alpha = 1, 2), \tag{2.7.22}$$

$$u^{(0)}_3(x_1, x_2) = 0 \quad \text{on } \partial D . \tag{2.7.23}$$

When $\langle f_\alpha \rangle = 0$, equilibrium equations (2.7.17) with constitutive equations (2.7.21) and boundary conditions (2.7.22) have the unique solution $u^{(0)}_\alpha(x_1, x_2) = 0$ $(\alpha = 1, 2)$ under the condition that the initial stresses do not cause stability loss of the membrane as a planar body. This condition is nearly always satisfied in practice since the initial stresses are small compared with the elasticity constants. Then, (2.7.20) takes the form,

$$N_{3\gamma x} = N^{*}_{\gamma\alpha} u^{(0)}_{3,\alpha x}(x_1, x_2), \tag{2.7.24}$$

with the equilibrium equation,

$$N_{3\gamma,\gamma x} + \langle f_3 \rangle + \langle g_3 \rangle_S = 0, \tag{2.7.25}$$

and boundary condition (2.7.23).

Equations (2.7.24) and (2.7.25) can be transformed into the classical equations for the flexure of a membrane,

$$(N^{*}_{\gamma\alpha} u^{(0)}_{3,\alpha x})_{,\gamma x} + \langle f_3 \rangle + \langle g_3 \rangle_S = 0.$$

The result obtained means that the elastic properties of a membrane cannot be neglected (as we neglect a small value as compared with a large value). We neglect not the elastic properties of a membrane, but the in-plate displacements, which are small compared with the normal deflection.

2.8 Plates with no Initial Stresses. Computing of Resultant Initial Stresses

The problem of a plate with no initial stresses is important in our consideration. In the previous sections, we introduced the resultant initial in-plane stresses by integrating the local stresses over the periodicity cell. This creates no problems for the theoretical analysis of stressed plates. However, this definition is not suitable for practical computations. In this section, we demonstrate that the resultant in-plane stresses, defined as above, are equal to the resultant in-plane stresses computed using a 2-D homogenization model of a plate. It means that an inhomogeneous plate (plate of complex structure) can be analyzed within the framework of 2-D models. This proposition cannot be extended to the problem of computing of homogenized stiffnesses, which is essentially a 3-D problem.

A homogenized model for a plate with no initial stresses

General homogenized models for plates with no initial stresses were obtained by Kohn and Vogelius (1984) and Caillerie (1984). The homogenized model of a plate with no initial stresses can be obtained from the model obtained above. It is sufficient to set $\sigma^{*(p)}_{ij} = 0$ (p=–3,–2) in (2.3.4) and the following formulas. We present here the more important formulas, which can be obtained in this way.

The governing equations of a 2-D plate are the following [see (2.2.45)–(2.2.47)]:

$$N^{(-2)}_{\gamma\delta} = A^{0}_{\gamma\delta\alpha\beta} V_{\alpha,\beta x} + A^{1}_{\gamma\delta\alpha\beta} u^{(0)}_{3,\alpha x\beta x}, \tag{2.8.1}$$

$$M^{(-2)}_{\gamma\delta} = A^{1}_{\gamma\delta\alpha\beta} V_{\alpha,\beta x} + A^{2}_{\gamma\delta\alpha\beta} u^{(0)}_{3,\alpha x\beta x}. \tag{2.8.2}$$

Here V_α ($\alpha = 1, 2$) and $u^{(0)}_3$ are determined from the solution of the homogenized 2-D plate problem.

The coefficients in (2.2.45) and (2.2.46) are

$$A^{\nu+\mu}_{\gamma\delta\alpha\beta} = \langle [(-1)^{\nu} c_{\gamma\delta\alpha\beta}(\mathbf{y}) y_3^{\nu} + c_{\gamma\delta mn}(\mathbf{y}) X^{\nu\alpha\beta}_{m,ny}(\mathbf{y})](-1)^{\mu} y_3^{\mu} \rangle \tag{2.8.3}$$

$$(\nu, \mu = 0, 1;\ \alpha, \beta, \gamma, \delta = 1, 2),$$

where $\mathbf{X}^{\nu\alpha\beta}(\mathbf{y})$ ($\nu = 0, 1$) is the solution of the cellular problem,

$$\begin{cases} (c_{ijmn}(\mathbf{y}) X^{\nu\alpha\beta}_{m,ny} + (-1)^{\nu} y_3^{\nu} c_{ij\beta\alpha}(\mathbf{y}))_{,jy} = 0 \quad \text{in } Y, \\ (c_{ijmn}(\mathbf{y}) X^{\nu\alpha\beta}_{m,ny} + (-1)^{\nu} y_3^{\nu} c_{ij\beta\alpha}(\mathbf{y})) n_j = 0 \quad \text{on S}, \\ \mathbf{X}^{\nu\alpha\beta}(\mathbf{y}) \text{ is periodic in } y_1, y_2 \text{ with periodicity cell } T. \end{cases}$$

The equilibrium equations are the following:

$$N^{(-2)}_{\alpha\beta,\beta x} + \langle f_\alpha \rangle + \langle g_\alpha \rangle_S = 0, \tag{2.8.4}$$

$$N^{(-1)}_{3\alpha,\alpha x} + \langle f_3 \rangle + \langle g_3 \rangle_S = 0, \tag{2.8.5}$$

$$M^{(-2)}_{\alpha\beta,\beta x} - N^{(-1)}_{\alpha 3} = 0, \tag{2.8.6}$$

$$(i, j = 1, 2, 3 \text{ and } \alpha, \beta = 1, 2).$$

In the case under consideration, $N^{(-1)}_{\alpha 3} = N^{(-1)}_{3\alpha}$, and from (2.8.5), (2.8.6) we obtain the equation,

$$M^{(-2)}_{\alpha\beta,\beta x\alpha x} + \langle f_3 \rangle + \langle g_3 \rangle_S = 0 \tag{2.8.7}$$

$(i,j$=1, 2, 3 and $\alpha, \beta = 1, 2$).

The following approximation for local stresses is valid [see (2.2.44)]:

$$\sigma^{(-2)}_{ij} \approx [c_{ijmn}(\mathbf{y}) X^{0\alpha\beta}_{m,ny}(\mathbf{y}) + c_{ij\alpha\beta}(\mathbf{y})] V_{\alpha,\beta x}(x_1, x_2) \tag{2.8.8}$$

$$+ [c_{ijmn}(\mathbf{y}) X^{1\alpha\beta}_{m,ny}(\mathbf{y}) - y_3 c_{ij\alpha\beta}(\mathbf{y})] u^{(0)}_{3,\alpha x\beta x}(x_1, x_2).$$

We write the equality (2.5.14) for the case $N^*_{\alpha\beta} = 0$. The quantity $E(0,\mathbf{u})$ is the elastic energy of a plate, considered as a 3-D elastic body. Thus,

$$E(0,\mathbf{u}) \approx E(u^{(0)}_3, V_1, V_2),$$

where $E(u^{(0)}_3, V_1, V_2)$ is the elastic energy computed in accordance with the 2-D homogenized model of a plate. This means that the elastic energy of a 2-D homogenized plate approximates the elastic energy of a 3-D plate (the elastic energy of a 3-D thin elastic structure).

Computing of resultant initial stresses and moments

A 2-D model of a plate with initial stresses involves the resultant in–plane initial stresses $N^*_{\alpha\beta}$ introduced by averaging the local stresses $\sigma^*_{\alpha\beta}$ [see (2.2.56)] over a 3-D periodicity cell Y:

$$N^*_{\alpha\beta} = (mesT)^{-1} \int_Y \sigma^*_{\alpha\beta}(x_1, x_2, \mathbf{y}) d\mathbf{y} = \langle \sigma^*_{\alpha\beta} \rangle, \tag{2.8.9}$$

where σ^*_{ij} is defined by solving the 3-D elasticity problem for a thin body of small diameter. The method does not look effective because we have to solve the 3-D elasticity problem using this method.

We prove that the resultant initial stresses and moments problem describing the initial state of a beam with no initial stresses can be determined by solving the 2-D problem with no initial stresses. This proof will give completeness to the developed theory because only 2-D problems will be involved in the model.

We consider the elasticity problem for a thin elastic layer (3-D plate) with no initial stresses. The problem can be analyzed using the method described above. The local stresses (in the case under consideration, they are the local initial stresses σ^*_{ij}) in a layer are given by (2.8.8):

$$\sigma^*_{ij} = [c_{ijmn}(\mathbf{y}) X^{0\alpha\beta}_{m,ny}(\mathbf{y}) + c_{ij\alpha\beta}(\mathbf{y})] V_{\alpha,\beta x}(x_1, x_2) \tag{2.8.10}$$

$$+ [c_{ijmn}(\mathbf{y}) X^{1\alpha\beta}_{m,ny}(\mathbf{y}) - y_3 c_{ij\alpha\beta}(\mathbf{y})] u^{(0)}_{3,\alpha x\beta x}(x_1, x_2) .$$

Substituting (2.8.10) in (2.8.9), we obtain the following formula:

$$N^*_{\alpha\beta} = \langle \sigma^*_{\alpha\beta} \rangle = \langle c_{ijmn}(\mathbf{y}) X^{0\alpha\beta}_{m,ny}(\mathbf{y}) + c_{ij\alpha\beta}(\mathbf{y}) \rangle V_{\alpha,\beta x}(x_1, x_2) \tag{2.8.11}$$

$$+ \langle c_{ijmn}(\mathbf{y}) X^{1\alpha\beta}_{m,ny}(\mathbf{y}) - y_3 c_{ij\alpha\beta}(\mathbf{y}) \rangle u^{(0)}_{3,\alpha x\beta x}(x_1, x_2) .$$

Here, V_α $(\alpha = 1,2), u^{(0)}_3$ is the solution of the 2-D plate problem for a plate with no initial stresses. In accordance with (2.2.47),

$$\langle c_{ijmn}(\mathbf{y}) X^{0\alpha\beta}_{m,ny}(\mathbf{y}) + c_{ij\alpha\beta}(\mathbf{y}) \rangle = A^0_{\alpha\beta\gamma\delta} , \tag{2.8.12}$$

$$\langle c_{ijmn}(\mathbf{y}) X^{1\alpha\beta}_{m,ny}(\mathbf{y}) - y_3 c_{ij\alpha\beta}(\mathbf{y}) \rangle = A^1_{\alpha\beta\gamma\delta} .$$

Using (2.8.12), we can write (2.8.11) as

$$N^*_{\alpha\beta} = \langle \sigma^*_{\alpha\beta} \rangle = A^0_{\alpha\beta\gamma\delta} V_{\alpha,\beta x}(x_1, x_2) + A^1_{\alpha\beta\gamma\delta} u^{(0)}_{3,\alpha x\beta x}(x_1, x_2) . \tag{2.8.13}$$

Comparing (2.8.12) with (2.2.45), we find that

$$N^*_{\alpha\beta} = N^{(-2)}_{\alpha\beta} , \tag{2.8.14}$$

where $N^*_{\alpha\beta}$ are the resultant in-plane stresses determined from the solution of the 2-D plate problem.

It seen that by averaging the local initial stresses, we will arrive at equality (2.8.14) only if equalities (2.8.12) are satisfied. In other words, we will arrive at equality (2.8.14) if the effective stiffnesses of the plate were computed in accordance with the asymptotic homogenization procedure.

Both methods described above (the averaging of the local stresses and the determination of the resultant stresses from the solution of the 2-D plate problem) incorporate the structure of a plate. The first method incorporates the structure of a plate in an explicit form. The second method incorporates the plate structure indirectly through the homogenized stiffnesses of the plate, $A^0_{\alpha\beta\gamma\delta}$ and $A^1_{\alpha\beta\gamma\delta}$, which depend on the structure of the plate.

3 Stressed Composite Beams, Rods, and Strings

This chapter is devoted to inhomogeneous stressed structures occupying small diameter regions (beam- and string-like structures). Theories based on 1-D and 3-D models are presented.

The homogenization models for periodically inhomogeneous beams were obtained in the 1970s using a 1-D beam model as the original model. The homogenized models obtained in this framework will be applicable only to those beams whose diameter is significantly less then the length of the period. This condition often is not fulfilled. Quite often, the diameter and the period of the beam are comparable in magnitude. Under these conditions, it is evident that a 3-D distribution of inhomogeneities should be taken into account, when one derives the homogenized 1-D model of an inhomogeneous beam. The homogenized models for beams considered as 3-D structures of small diameter were developed in the 1980s (after homogenized models for composite solids and plates).

The homogenization models for inhomogeneous beams of coaxial structure can be found in the works by Trabucho and Viaño (1987), Tutek and Aganovich (1987), and Kozlova (1989) (see the bibliography in the book by Trabucho and Viaño, 1996). Kolpakov (1991) analyzed a beam of arbitrary periodic structure. In the papers mentioned, beams with no initial stresses were considered. The transition from the 3-D elasticity problem in a thin domain to the 1-D beam model for stressed structures was made in the paper by Kolpakov (1992a).

3.1 Stressed Beam (1-D Model)

As noted above, for beams whose diameter is significantly less than the length of the period, 1-D equations of the theory of thin inhomogeneous beams may be taken as the initial model. In this section, we briefly outline the asymptotic expansions method as applied to the 1-D model of an inhomogeneous beam with periodic structure.

Formulation of the problem

We will examine a beam of periodic structure obtained by repeating a certain small periodicity cell $\varepsilon[0,T]$. Here ε is the periodicity cell's characteristic dimension, which is assumed to be small (that is formalized in the form $\varepsilon \to 0$).

The starting point of our investigation is the 1-D formulation of a beam problem for a beam with initial stresses. In accordance with this model, the equilibrium equations of the beam with initial stresses can be written in the form,

$$\frac{d^2}{dx^2}\left(\frac{d^2 M^\varepsilon}{dx^2}+\sigma^*\frac{d^2 w^\varepsilon}{dx^2}\right)=f(x,x/\varepsilon)\text{ in }[-a,a], \tag{3.1.1}$$

where

$$M^\varepsilon = D(x/\varepsilon)\frac{d^2 w^\varepsilon}{dx^2}, \tag{3.1.2}$$

$D(x/\varepsilon)$ is the beam stiffness calculated using the plane cross section hypothesis, M^ε is the moment, σ^* is the initial axial force calculated using 1-D rod model, and $f(x,x/\varepsilon)$ is the normal force (the sum of mass and surface forces).

If axial stress σ^* satisfies the equation $d\sigma^*/dx=0$, then $\sigma^*=const$. This means that the initial axial force does not depend on the "fast" variable.

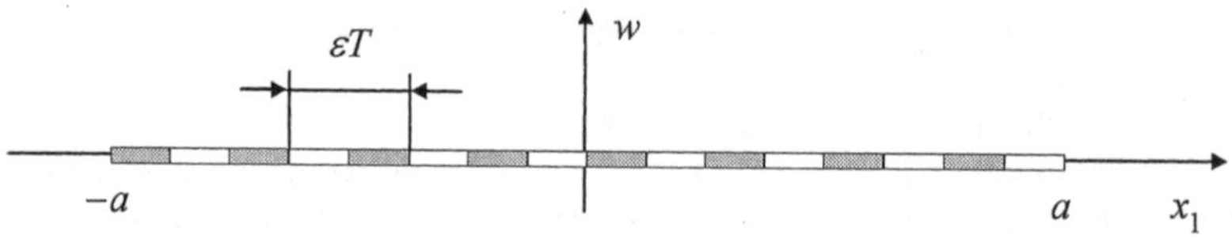

Fig. 3.1. Beam of a periodic structure

The boundary conditions can be written in the form,

$$w^\varepsilon(-a)=w^\varepsilon(a)=\frac{dw^\varepsilon}{dx}(-a)=\frac{dw^\varepsilon}{dx}(a)=0. \tag{3.1.3}$$

Here, $-a$ and a are coordinates of the ends of the beam, the beam is clamped on the ends (see Fig. 3.1), and w^ε is the normal deflection of the beam.

Asymptotic expansions and equations of equilibrium

Using the two-scale method (see Sect. 1.2), a function $f(x,x/\varepsilon)$ of the arguments x and x/ε is considered a function $f(x,y)$ of "slow" variable x and "fast" variable $y=x/\varepsilon$. In accordance with this note, the function $D(x/\varepsilon)$ will be written in the form $D(y)$. The function $D(y)$ is periodic in y with period T corresponding to the period of the beam structure, see Fig. 3.1.

We will study global deformations of the beam, as $\varepsilon\to 0$. To do this, we use the following asymptotic expansion: asymptotic expansion for deflection,

$$w^{\varepsilon} = w^{(0)}(x) + \varepsilon^2 w^{(2)}(x, y) + \ldots = w^{(0)}(x) + \sum_{k=2}^{\infty} \varepsilon^k w^{(k)}(x,y), \tag{3.1.4}$$

asymptotic expansion for moments,

$$M_{ij} = \sum_{k=0}^{\infty} \varepsilon^k M_{ij}^{(k)}(x,y). \tag{3.1.5}$$

Note that in expansion (3.1.4) for deflection, the term $w^{(0)}(x)$ depends only on the variables x and the term of the order of ε is taken equal to zero.

The differential operators are presented in the form of the sum of operators in x and in y (see Chap. 1). This representation for the function $Z(x,y)$ of the arguments x and y, as in the right-hand sides of (3.1.4), (3.1.5), takes the form,

$$\frac{dZ}{dx_i} = Z_{,x} + \varepsilon^{-1} Z_{,y}, \tag{3.1.6}$$

$$\frac{d^2 Z}{dx^2} = Z_{,xx} + \varepsilon^{-1}(Z_{,xy} + Z_{,yx}) + \varepsilon^{-2} Z_{,yy}.$$

Substituting (3.1.4), (3.1.5) in (3.1.1) and replacing the differential operators in accordance with the differentiating rule (3.1.6), we obtain

$$\sum_{k=0}^{\infty} \varepsilon^k [M_{,xx}^{(k)} + \varepsilon^{-1}(M_{,xy}^{(k)} + M_{,yx}^{(k)}) + \varepsilon^{-2} M_{,yy}^{(k)}] \tag{3.1.7}$$

$$= \sum_{k=0}^{\infty} \varepsilon^k [\sigma_{ij}^{*}(w_{,x}^{(k)} + \varepsilon^{-1} w_{,y}^{(k)}))_{,x} + \varepsilon^{-1}(\sigma_{ij}^{*}(w_{,x}^{(k)} + \varepsilon^{-1} w_{,y}^{(k)})]_{,y}.$$

Proposition 3.1. *The following relationship holds: for* $\varepsilon \to 0$,

$$\int_{-a}^{a} Z(x, x/\varepsilon)\mathrm{d}x = \int_{-a}^{a} \langle Z\rangle(x)\mathrm{d}x$$

for any function $Z(x,y)$ *periodic in y with period* T.

Here,

$$\langle \circ \rangle = T^{-1}\int_0^T \circ\, dy$$

is the average value over the period T in "fast" variable y (Fig. 3.1).

Averaging (3.1.7), we obtain

$$\sum_{k=0}^{\infty} \varepsilon^{k-2} \langle M^{(k)} \rangle_{,xx} + \sum_{k=0}^{\infty} \varepsilon^{k} \sigma^{*} w^{(k)}_{,xx} = \langle f \rangle (x) . \tag{3.1.8}$$

Here we use the equality $\langle Z_{,y} \rangle = 0$, which is fulfilled for every function $Z(y)$ periodic with period T. This equation follows from the equation,

$$\int_0^T Z_{,y}(y)dy = Z(T) - Z(0) ,$$

and periodicity condition $Z(T) = Z(0)$.

Equating terms of the same order of ε in (3.1.9), we obtain an infinite sequence of the homogenized equation of equilibrium, the first of which has the following form:

$$\langle M^{(0)} \rangle_{,xx} + \sigma^{*} w^{(0)}_{,xx} = \langle f \rangle (x) . \tag{3.1.9}$$

Equating terms of the same order of ε in (3.1.7), we obtain an infinite sequence of the local equations of equilibrium, the first of which has the following form :

$$M^{(0)}_{,yy} = 0 . \tag{3.1.10}$$

Substituting (3.1.4), (3.1.5) in (3.1.2) and replacing the differential operators in accordance with (3.1.6),

$$\sum_{k=0}^{\infty} \varepsilon^{k} M^{(k)} = \sum_{k=0}^{\infty} \varepsilon^{k} [D(y)(w^{(k)}_{,xx} + \varepsilon^{-1}(w^{(k)}_{,xy} + w^{(k)}_{,yx}) + \varepsilon^{-2} w^{(k)}_{,yy})] . \tag{3.1.11}$$

Equating terms of the same order of ε in (3.1.11), we obtain

$$M^{(0)} = D(y) w^{(2)}_{,yy} + D(y) w^{(0)}_{,xx}(x) . \tag{3.1.12}$$

Substituting (3.1.12) in (3.1.10),

$$[D(y) w^{(2)}_{,yy} + D(y) w^{(0)}_{,xx}(x)]_{,yy} = 0 . \tag{3.1.13}$$

The periodic solution of problem (3.1.13) is given by the formula,

$$w^{(2)} = X(y)w^{(0)}_{,xx}(x) + V(x), \tag{3.1.14}$$

where $X(y)$ is the solution of the following cellular problem:

$$\begin{cases} [D(y)X_{,yy} + D(y)]_{,yy} = 0, \ x \in [0,T], \\ X(y) \text{ and } X_{,y}(y) \text{ are periodic in } y \text{ with period } T. \end{cases} \tag{3.1.15}$$

Substituting (3.1.14) in (3.1.12), we obtain

$$M^{(0)}_{ij} = [D(y)X_{,yy}(y) + D(y)]w^{(0)}_{,xx}(x). \tag{3.1.16}$$

Averaging (3.1.16), we obtain the homogenized governing equation,

$$\langle M^{(0)} \rangle = \mathsf{D}w^{(0)}_{,xx}(x), \tag{3.1.17}$$

where

$$\mathsf{D} = \langle D(y)X_{,yy}(x) + D(y) \rangle \tag{3.1.18}$$

is the homogenized bending stiffness of the beam.

Boundary conditions are derived from the initial boundary conditions (3.1.2) and asymptotic expansion (3.1.4) as follows:

$$w(-a) = w(a) = \frac{dw}{dx}(-a) = \frac{dw}{dx}(a) = 0 . \tag{3.1.19}$$

The homogenized equation of equilibrium (3.1.9) with allowance for the homogenized governing equation (3.1.17) can be written as

$$(\mathsf{D}w_{,xx})_{,xx} + \sigma^* w^{(0)}_{,xx} = \langle f \rangle(x). \tag{3.1.20}$$

3.2 Stressed Beams (Transition from 3-D to 1-D Model, the Axial Initial Stresses)

In this section, we will consider non–homogeneous beams of periodic structure (which are widely used in modern structures). The non–homogeneity of a beam can be a result of the nonhomogeneity of the material from which the beam is made, and nonhomogeneity of the beam geometry. Both cases will be covered by our consideration. The classical uniform beam is a partial case of the beams under consideration.

The analysis of a small-diameter stressed composite body differs from both the analysis of 3-D composites with initial stresses given by Kolpakov (1989) and the analysis of thin beams with no initial stresses given in the paper by Kolpakov (1991). The difference between the given problem and the problems examined in the works considering beams with no initial stresses is related to the asymmetry of the coefficients; this requires a more detailed analysis of the problem. The difference between the given problem and the problem of a stressed 3-D body is connected with the differences in asymptotic expansions for monolithic body, plate, and beam.

Torsion of a stressed homogeneous cylindrical beam was studied earlier (see, e.g., Washizu, 1982) on the basis of the Saint-Venant theory. It is clear that the Saint-Venant theory cannot be used for beams of arbitrary periodic structure.

As will be seen from the discussion below, the order of the initial stresses relative to the characteristic diameter of the beam ε, plays a significant role in the problem. To account for this, we take the initial stresses σ_{ij}^{*} in the form $\sigma_{ij}^{*(p)} = \varepsilon^{-3}\sigma_{ij}^{*(-3)} + \varepsilon^{-2}\sigma_{ij}^{*(-2)}$.

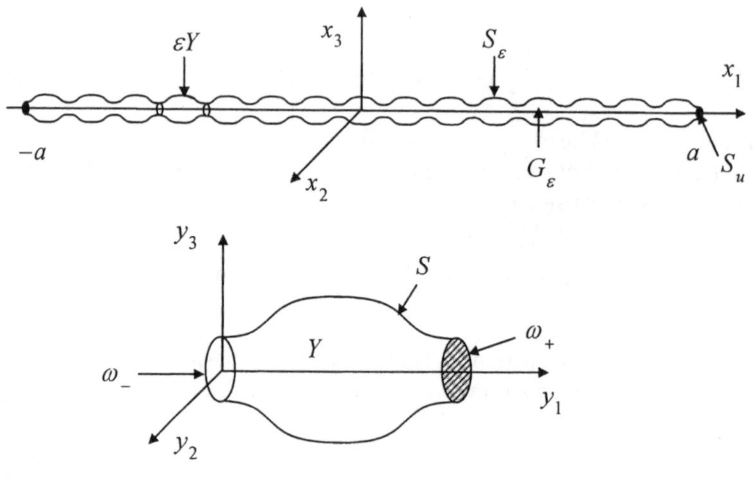

Fig. 3.2. A beam of periodic structure and its periodicity cell in "fast" variables

Formulation of the problem

We will examine a body of periodic structure obtained by repeating a certain small periodicity cell (periodicity cell) εY along the Ox_1 axis (Fig. 3.2). Here, ε is the periodicity cell's characteristic dimension, which is assumed to be small (that is formalized in the form, $\varepsilon \to 0$). As a result, we have a body of periodic structure with a small diameter - a 3-D beam. For $\varepsilon \to 0$, the 3-D beam "tightens" to the segment $[-a,a]$ at the Ox_1 axis (Fig. 3.2) - a 1-D beam. Our aim is to derive a model describing a 1-D beam.

The starting point of our investigation is the exact 3-D formulation of an elasticity problem for a body with initial stresses without any simplifying assumptions, presented in Sect. 1.1. In accordance with this model, the equilibrium equations of the beam as a 3-D elastic body with initial stresses can be written in the form,

$$\frac{\partial}{\partial x_j}[h_{ijmn}(x_1,\mathbf{x}/\varepsilon)\frac{\partial u_m^\varepsilon}{\partial x_n}]+ f_i^\varepsilon(\mathbf{x})=0 \text{ in } G_\varepsilon . \tag{3.2.1}$$

The mass force is taken in the form $\mathbf{f}^\varepsilon(\mathbf{x}) = (\varepsilon^{-2}f_1, \varepsilon^{-2}f_2, \varepsilon^{-3}f_3)\ (x_1,\mathbf{x}/\varepsilon)$.

The boundary conditions can be written in the form

$$h_{ijmn}(x_1,\mathbf{x}/\varepsilon)\frac{\partial u_m^\varepsilon}{\partial x_n}n_j = g_i^\varepsilon(\mathbf{x}) \quad \text{on } S_\varepsilon, \tag{3.2.2}$$

$$\mathbf{u}^\varepsilon(\mathbf{x})=0 \quad \text{on } S_u .$$

The surface force is taken in the form $\mathbf{g}^\varepsilon(\mathbf{x}) = (\varepsilon^{-1}g_1, \varepsilon^{-1}g_2, \varepsilon^{-2}\ g_3)(x_1,\mathbf{x}/\varepsilon)$.

Here, G_ε is the small diameter region occupied by the beam; S_ε is the lateral surface of the beam, $\mathbf{n}$ is a normal to S_ε; the beam is fastened along the surface S_u (Fig. 3.2); and $\mathbf{u}^\varepsilon$ are the displacements. The coefficients h_{ijmn} are a known (see Sect. 1.1) combination of the tensor of elasticity constants $\varepsilon^{-4}c_{ijmn}$ and initial stresses σ_{ij}^*, which, in the case under consideration, is taken in the form,

$$h_{ijmn}(x_1,\mathbf{x}/\varepsilon)=\varepsilon^{-4}c_{ijmn}(\mathbf{x}/\varepsilon)+\varepsilon^{-3}b_{ijmn}^{(-3)}(x_1,\mathbf{x}/\varepsilon)+\varepsilon^{-2}b_{ijmn}^{(-2)}(x_1,\mathbf{x}/\varepsilon), \tag{3.2.3}$$

where

$$b_{ijmn}^{(p)}(x_1,\mathbf{x}/\varepsilon)=\sigma_{jn}^{*(p)}(x_1,\mathbf{x}/\varepsilon)\delta_{im}\ (p=-3,-2). \tag{3.2.4}$$

Using the two-scale method, a function $f(x_1,\mathbf{x}/\varepsilon)$ of the arguments x_1 and $\mathbf{x}/\varepsilon$ is considered a function $f(x_1,\mathbf{y})$ of one "slow" variable x_1 and "fast" variables $\mathbf{y}=\mathbf{x}/\varepsilon$. In accordance with this note, functions $c_{ijmn}(\mathbf{x}/\varepsilon)$, $\sigma_{jn}^{*(p)}(x_1,\mathbf{x}/\varepsilon)$ will be written in the form, $c_{ijmn}(\mathbf{y})$, $\sigma_{jn}^{*(p)}(x_1,\mathbf{y})$. Functions $c_{ijmn}(\mathbf{y})$, $\sigma_{jn}^{*(p)}(x_1,\mathbf{y})$ are periodic in y_1 with period T corresponding to the period of the beam structure (T is the projection of periodicity cell Y on the Oy_1 axis in "fast" variables, see Fig. 3.2).

Stresses σ_{ij} determined by the formula,

$$\sigma_{ij} = \varepsilon^{-4} h_{ijmn}(\mathbf{x}/\varepsilon) \frac{\partial u_m^\varepsilon}{\partial x_n}, \tag{3.2.5}$$

are additional stresses (see Sect. 1.1). Formula (3.2.5) can be considered a local governing equation for the body under consideration.

In connection with the fact that the coefficients (3.2.5) are written in a form different from that normally used, we will comment briefly on the terms in (3.2.5). Tensor $\varepsilon^{-4} c_{ijmn}(\mathbf{x}/\varepsilon)$ describes the elastic constants of the material from which the beam is made. Multiplier ε^{-4} guarantees that the bending stiffnesses of the beam will be non–zero, as $\varepsilon \to 0$. The other terms describe the initial stresses. The term $\varepsilon^{-2} b_{ijmn}^{(-2)} = \varepsilon^{-2} \sigma_{jn}^{*(-2)} \delta_{im}$ corresponds to resultant forces, which are independent of ε. In fact, the resultant forces are equal to the stress multiplied by the cross-sectional area, which has a characteristic value equal to ε^{-2}. The condition that the resultant axial force corresponding to $\sigma_{jn}^{*(-3)}$ is not equal to zero is written in the form,

$$\int_Y \sigma_{11}^{*(-2)} \mathrm{d}\mathbf{y} \neq 0. \tag{3.2.6}$$

The term $\varepsilon^{-3} b_{ijmn}^{(-3)} = \varepsilon^{-3} \sigma_{jn}^{*(-3)} \delta_{im}$ corresponds to moments, which are independent of ε.

Asymptotic expansion and equations of equilibrium

We will study global deformations of the beam, in particular, buckling in global forms, as $\varepsilon \to 0$. To do this, we use the following asymptotic expansion proposed in the work by Kolpakov (1991): asymptotic expansion for displacements,

$$\mathbf{u}^\varepsilon = \mathbf{u}^{(0)}(x_1) + \varepsilon \mathbf{u}^{(1)}(x_1, \mathbf{y}) + \ldots = \mathbf{u}^{(0)}(x_1) + \sum_{k=1}^{\infty} \varepsilon^k \mathbf{u}^{(k)}(x_1, \mathbf{y}), \tag{3.2.7}$$

asymptotic expansion for stresses,

$$\sigma_{ij} = \sum_{p=-4}^{\infty} \varepsilon^p \sigma_{ij}^{(p)}(x_1, \mathbf{y}). \tag{3.2.8}$$

Here x_1 is a "slow" variable along the axis of the beam $[-a, a]$; $\mathbf{y} = \mathbf{x}/\varepsilon$ are "fast" variables. The functions on the right-hand side of (3.2.7), (3.2.8) are assumed to be periodic in y_1 with period T. Note that the term $\mathbf{u}^{(0)}(x_1)$ in (3.2.7) depends only on "slow" variable x_1. Expansion (3.2.8) starts with a term of order ε^{-4} in accordance with governing equation (3.2.5) and expansion (3.2.7) for displacements.

The analysis of problem (3.2.1)–(3.2.4) breaks down into two stages. The first entails obtaining the equations of equilibrium for the beam considered as a 1-D structure. As in the case of plates (see Chap. 2), this stage is not involved with local governing equations [in the case under consideration, with the equation (3.2.5)], and it is the same stage for any governing equations.

Using the two-scale expansion, the differential operators are presented in the form of the sum of operators in $\{x_i\}$ and in $\{y_i\}$. For the function $Z(x_1, \mathbf{y})$ of the arguments x_1 and $\mathbf{y} = (y_1, y_2, y_3)$, as in the right-hand sides of (3.2.7), (3.2.8), this representation takes the form,

$$\frac{\partial Z}{\partial x_1} = Z_{,1x} + \varepsilon^{-1} Z_{,1y}, \tag{3.2.9}$$

$$\frac{\partial Z}{\partial x_\alpha} = \varepsilon^{-1} Z_{,\alpha y} \ (\alpha = 2, 3).$$

Here and below in this chapter, the Latin indexes take the values 1,2,3 and the Greek indexes take the values 2,3; subscript $,\alpha x$ means $\partial / \partial x_\alpha$, and subscript $,1y$ means $\partial / \partial y_1$.

We introduce

$$\langle \circ \rangle = T^{-1} \int_Y \circ \, d\mathbf{y} \text{ and } \langle \circ \rangle_S = T^{-1} \int_S \circ \, d\mathbf{y}, \tag{3.2.10}$$

the average values over periodicity cell Y and its lateral surface S (Fig.3.2).

Proposition 3.2. *The following relationships holds: as* $\varepsilon \to 0$,

$$\varepsilon^{-2} \int_{G_\varepsilon} Z(x_1, \mathbf{x}/\varepsilon) d\mathbf{x} \to \int_{-a}^{a} \langle Z \rangle (x_1) dx_1,$$

$$\varepsilon^{-1} \int_{S_\varepsilon} Z(x_1, \mathbf{x}/\varepsilon) d\mathbf{x} \to \int_{-a}^{a} \langle Z \rangle_S (x_1) dx_1,$$

for every function $Z(x_1, \mathbf{y})$ *periodic in* y_1 *with period* T.

Allowing for (3.2.5), problem (3.2.1), (3.2.2.) can be written in the form,

$$\frac{\partial \sigma_{ij}}{\partial x_j} + f_i^{\varepsilon}(\mathbf{x}) = 0 \quad \text{in } G_{\varepsilon}, \; \sigma_{ij} n_j = g_i^{\varepsilon}(\mathbf{x}) \quad \text{on } S_{\varepsilon}. \tag{3.2.11}$$

In order to obtain the homogenized equilibrium equations, we introduce the functional space (the set of admissible displacements, in other words),

$$U_{\varepsilon} = \{\mathbf{v} \in \{H^1(G_{\varepsilon})\}^3 : \mathbf{v}(\mathbf{x}) = 0 \quad \text{on } S_u\},$$

and consider a weak form of the local equilibrium equation (3.2.11):

$$\int_{G_{\varepsilon}} \sigma_{ij} \frac{\partial v_i}{\partial x_j} d\mathbf{x} = \int_{S_{\varepsilon}} \mathbf{g}^{\varepsilon} \mathbf{v} d\mathbf{x} + \int_{G_{\varepsilon}} \mathbf{f}^{\varepsilon} \mathbf{v} d\mathbf{x} \quad \text{for any } \mathbf{v} \in U_{\varepsilon}. \tag{3.2.12}$$

Substituting expansion (3.2.8) in the equilibrium equations (3.2.12), we obtain, with allowance for (3.2.9),

$$\sum_{p=-4}^{\infty} \int_{G_{\varepsilon}} \varepsilon^{p+2} \sigma_{ij}^{(p)} (v_{i,1x} \delta_{j1} + \varepsilon^{-1} v_{i,jy}) d\mathbf{x} \tag{3.2.13}$$

$$= \int_{S_{\varepsilon}} \mathbf{g}^{\varepsilon} \mathbf{v} d\mathbf{x} + \int_{G_{\varepsilon}} \mathbf{f}^{\varepsilon} \mathbf{v} d\mathbf{x} \quad \text{for any } \mathbf{v} \in U_{\varepsilon}.$$

We introduce the functional space,

$$H = \{\mathbf{w}(x_1) \in \{H^1([-a, a])\}^3 : \mathbf{w}(-a) = \mathbf{w}(a) = 0\}.$$

It is obvious that $H \subset U_{\varepsilon}$.

We set $\mathbf{v} = \mathbf{w}(x_1) \in H$ in (3.2.13). Then, (3.2.13) takes the following form:

$$\sum_{p=-4}^{\infty} \int_{G_{\varepsilon}} \varepsilon^{p} \sigma_{i1}^{(p)} w_{i,1x} d\mathbf{x} = \int_{S_{\varepsilon}} \mathbf{g}^{\varepsilon} \mathbf{w} d\mathbf{x} + \int_{G_{\varepsilon}} \mathbf{f}^{\varepsilon} \mathbf{w} d\mathbf{x} \quad \text{for any } \mathbf{w} \in H.$$

By virtue of Proposition 3.2 and the definition of forces $\mathbf{f}^{\varepsilon}$ and $\mathbf{g}^{\varepsilon}$ [see (3.2.1) and (3.2.2)], we find from this equation that

$$\sum_{p=-4}^{\infty} \int_{-a}^{a} \varepsilon^{p+2} \langle \sigma_{i1}^{(p)} \rangle w_{i,1x} \mathrm{d}x_1 \tag{3.2.14}$$

$$= \int_{-a}^{a} (\langle g_\alpha \rangle_S w_\alpha + \varepsilon^{-1} \langle g_1 \rangle_S w_1) \mathrm{d}x_1$$

$$+ \int_{-a}^{a} (\langle f_\alpha \rangle w_\alpha + \varepsilon^{-1} \langle f_1 \rangle w_1) \mathrm{d}x_1 \quad \text{for any } \mathbf{w} \in H .$$

We introduce the resultant stresses by the formula (note that the integration is over periodicity cell Y, not over a cross section),

$$N_{ij}^{(p)}(x_1) = \langle \sigma_{ij}^{(p)}(x_1, \mathbf{y}) \rangle$$

$$(i, j = 1, 2, 3 \text{; and } p = -4, -3, \ldots).$$

Then, we can write (3.2.14) as

$$\sum_{p=-4}^{\infty} \int_{-a}^{a} \varepsilon^{p+2} N_{i1}^{(p)} w_{i,1x} \mathrm{d}x_1$$

$$= \int_{-a}^{a} (\langle g_\alpha \rangle_S w_\alpha + \varepsilon^{-1} \langle g_1 \rangle_S w_1) \mathrm{d}x_1$$

$$+ \int_{-a}^{a} (\langle f_\alpha \rangle w_\alpha + \varepsilon^{-1} \langle f_1 \rangle w_1) \mathrm{d}x_1 \quad \text{for any } \mathbf{w} \in H .$$

Integrating by parts, we obtain the following set of equilibrium equations with respect to the resultant stresses:

$$N_{11,1x}^{(-4)} = 0 , \tag{3.2.15}$$

$$N_{11,1x}^{(-3)} + \langle f_1 \rangle + \langle g_1 \rangle_S = 0 ,$$

$$N_{\beta 1,1x}^{(-2)} + \langle f_\beta \rangle + \langle g_\beta \rangle_S = 0 . \tag{3.2.16}$$

The equations with $p > 2$ are not involved in our analysis.

We set $\mathbf{v} = \mathbf{w}(x_1)y_\alpha$, where $\phi \in H$, in (3.2.13) (it is clear that $\mathbf{w}(x_1)y_\alpha \in U_\varepsilon$). Then, (3.2.13) takes the following form:

$$\sum_{p=-4}^{\infty} \int_{G_\varepsilon} \varepsilon^p \sigma_{ij}^{(p)} (w_{i,1x}\delta_{j1} y_\alpha + \varepsilon^{-1} w_i \delta_{j\alpha}) d\mathbf{x}$$

$$= \int_{S_\sigma} \mathbf{g}^\varepsilon y_\alpha \mathbf{w} d\mathbf{x} + \int_{G_\varepsilon} \mathbf{f}^\varepsilon y_\alpha \mathbf{w} d\mathbf{x} \quad \text{for any } \mathbf{w} \in H.$$

By virtue of Proposition 3.2, we find from this equation that

$$\sum_{p=-4}^{\infty} \int_{-a}^{a} \varepsilon^{p+2} (\langle \sigma_{i1}^{(p)} y_\alpha \rangle w_{i,1x} + \varepsilon^{-1} \langle \sigma_{i\alpha}^{(p)} \rangle w_i) dx_1 \tag{3.2.17}$$

$$= \int_{S_\sigma} (\langle g_\chi y_\alpha \rangle_S w_\chi + \varepsilon^{-1} \langle g_1 y_\alpha \rangle_S w_1) d\mathbf{x}$$

$$+ \int_{G_\varepsilon} (\langle f_\chi y_\alpha \rangle w_\chi + \varepsilon^{-1} \langle f_1 y_\alpha \rangle w_1) d\mathbf{x} \quad \text{for any } \mathbf{w} \in H.$$

We introduce the resultant moments by the formula (the integration is over periodicity cell Y, not over a cross section),

$$M_{i\beta}^{(p)}(x_1) = \langle \sigma_{j1}^{(p)}(x_1, \mathbf{y}) y_\beta \rangle \quad (i, j = 1, 2, 3;\ \beta = 2, 3 \text{ and } p = -4, -3, \ldots).$$

Then, (3.2.17) can be written as

$$\sum_{p=-4}^{\infty} \int_{-a}^{a} \varepsilon^{p+2} (M_{i\alpha}^{(p)} w_{i,1x} + \varepsilon^{-1} N_{i\alpha}^{(p)} w_i) dx_1$$

$$= \int_{S_\sigma} (\langle g_\chi y_\alpha \rangle_S w_\chi + \varepsilon^{-1} \langle g_1 y_\alpha \rangle_S w_1) d\mathbf{x}$$

$$+ \int_{G_\varepsilon} (\langle f_\chi y_\alpha \rangle w_\chi + \varepsilon^{-1} \langle f_1 y_\alpha \rangle w_1) d\mathbf{x} \quad \text{for any } \mathbf{w} \in H.$$

Integrating by parts, we obtain the following set of equilibrium equations:

$$M^{(-4)}_{i\beta,1x} - N^{(-3)}_{i\beta} = 0\,, \tag{3.2.18}$$

$$M^{(-3)}_{i\beta,1x} - N^{(-2)}_{i\beta} + \delta_{i1}(\langle f_1 y_3\rangle + \langle g_1 y_3\rangle_S) = 0\,,$$

$$\langle \sigma^{(-3)}_{i\alpha}\rangle = 0\,. \tag{3.2.19}$$

Equations (3.2.18) can be broken down into the following two groups: equilibrium equations for the bending moments $M^{(-3)}_{1\beta}$,

$$M^{(-3)}_{1\beta,1x} - N^{(-2)}_{1\beta} + \langle f_1 y_3\rangle + \langle g_1 y_3\rangle_S = 0\,, \tag{3.2.20}$$

and equilibrium equation for the torsional moment $M = M^{(-3)}_{32} - M^{(-3)}_{23}$,

$$M_{,1x} + (N^{(-2)}_{32} - N^{(-2)}_{23}) = 0\,. \tag{3.2.21}$$

Equations (3.2.15), (3.2.16), (3.2.20), and (3.2.21) look like the classical ones. But in contrast to the classical equations in (3.2.16), (3.2.20),

$$N^{(-2)}_{\beta 1} \neq N^{(-2)}_{1\beta}$$

and in (3.2.21)

$$N^{(-2)}_{23} \neq N^{(-2)}_{32}\,.$$

Substituting expansion (3.2.8) in the equilibrium equations (3.2.11), we obtain, with allowance for (3.2.9),

$$\sum_{p=-4}^{\infty}(\varepsilon^p \sigma^{(p)}_{i1,1x} + \varepsilon^{p-1}\sigma^{(p)}_{ij,jy}) + f_i(x_1,\mathbf{y}) = 0 \quad \text{in } G_\varepsilon\,, \tag{3.2.22}$$

$$\sum_{p=-4}^{\infty}\varepsilon^p \sigma^{(p)}_{ij,jy} = g^{\varepsilon}_i(\mathbf{x}) \quad \text{on } S_\varepsilon\,.$$

Equating the terms with identical power in ε (3.2.22), we obtain the following equations:

$$\sigma^{(p)}_{ij,jy} + \sigma^{(p-1)}_{i1,1x} = 0 \quad \text{in } Y\,,$$

$$\sigma^{(p)}_{ij} n_j = 0 \quad \text{on } S \;\; (p = -4, -3, \ldots), \tag{3.2.23}$$

$$\sigma_{ij}^{(-5)} n_j = 0 .$$

Here S means the lateral surface of periodicity cell Y in "fast" variables; **n** is the normal to S (see Fig. 3.2).

Note that the equilibrium equations (3.2.15), (3.2.16), (3.2.20), and (3.2.21) with respect to the resultant stresses and moments are independent of the local governing relations.

Homogenized governing equations

The second stage of the analysis of the problem consists of obtaining the governing equations for the beam as a 1-D structure and excluding unknown quantities from the equilibrium equations. In contrast to the first stage, this stage does involve local governing equations.

In this section, we analyze the case,

$$\sigma_{ij}^{*(-3)} = 0 \text{ and } \sigma_{11}^{*(-2)} \neq 0 ,$$

corresponding to nonzero axial force.

Furthermore, in this section, $\overset{*}{\sigma}_{ij}$ means $\sigma_{ij}^{*(-2)}$, and b_{ijkl} means $b_{ijkl}^{(-2)}$ [coefficients $b_{ijkl}^{(p)}$ are determined by (3.2.4)].

Substituting (3.2.7), (3.2.8) in local governing equation (3.2.5), with allowance for (3.2.9), we obtain

$$\sum_{p=-4}^{\infty} \varepsilon^p \sigma_{ij}^{(p)} \tag{3.2.24}$$

$$= \sum_{k=0}^{\infty} \varepsilon^k [\varepsilon^{-4} c_{ijmn}(\mathbf{y}) + \varepsilon^{-2} b_{ijmn}(x_1, \mathbf{y})][u_{m,1x}^{(k)} \delta_{n1} + \varepsilon^{-1} u_{m,ny}^{(k)}] .$$

Equating the terms with identical power of ε in (3.2.24), we obtain

$$\sigma_{ij}^{(p)} = c_{ijm1}(\mathbf{y}) u_{m,1x}^{(p+4)} + c_{ijmn}(\mathbf{y}) u_{m,ny}^{(p+5)} \quad (p = -4, -3, \dots), \tag{3.2.25}$$

$$\sigma_{ij}^{(-2)} = c_{ijm1}(\mathbf{y}) u_{m,1x}^{(2)} + b_{ijm1}(x_1, \mathbf{y}) u_{m,1x}^{(0)} \tag{3.2.26}$$

$$+ c_{ijmn}(\mathbf{y}) u_{m,ny}^{(3)} + b_{ijmn}(x_1, \mathbf{y}) u_{m,ny}^{(1)} .$$

As was assumed above,

$$\mathbf{u}^{(k)}(\mathbf{y}) \text{ is periodic in } y_1 \text{ with period } T, \quad k = 1, 2, \ldots \tag{3.2.27}$$

Let us consider problem (3.2.23, $p = -4$), (3.2.25, $p = -4$), (3.2.27, $k = 1$). This problem can be written as

$$\begin{cases} [c_{ijmn}(\mathbf{y})u^{(1)}_{m,ny} + c_{ijm1}(\mathbf{y})u^{(0)}_{m,1x}(x_1)]_{,jy} = 0 & \text{in } Y, \\ [c_{ijmn}(\mathbf{y})u^{(1)}_{m,ny} + c_{ijm1}(\mathbf{y})u^{(0)}_{m,1x}(x_1)]n_j = 0 & \text{on } S, \\ \mathbf{u}^{(1)}(\mathbf{y}) \text{ is periodic in } y_1 \text{ with period } T. \end{cases} \tag{3.2.28}$$

In order to solve problems of this kind, cellular problems are introduced. In the case under consideration, we introduce the functions $\mathbf{X}^{0p}(\mathbf{y})\,(p = 1, 2, 3)$, which are determined by solutions of cellular problems of the beam theory (corresponding to axial deformation):

$$\begin{cases} [c_{ijmn}(\mathbf{y})X^{0p}_{m,ny} + c_{ijp1}(\mathbf{y})]_{,jy} = 0 & \text{in } Y, \\ [c_{ijmn}(\mathbf{y})X^{0p}_{m,ny} + c_{ijp1}(\mathbf{y})]n_j = 0 & \text{on } S, \\ \mathbf{X}^{0p}(\mathbf{y}) \text{ is periodic in } y_1 \text{ with period } T. \end{cases} \tag{3.2.29}$$

Proposition 3.3. *Solutions of cellular problem* (3.2.29) *for* $m = \alpha = 2, 3$ *are the following*:

$$\mathbf{X}^{0\alpha}(\mathbf{y}) = -y_\alpha \mathbf{e}_1 . \tag{3.2.30}$$

Here, $\{\mathbf{e}_i\}$ are basis vectors of the coordinate system.

To verify equality (3.2.30), let us substitute it in (3.2.29). We obtain $c_{ijmn}(\mathbf{y})X^{0\alpha}_{m,ny}(\mathbf{y}) + c_{ij\alpha 1}(\mathbf{y}) = -c_{ij1\alpha}(\mathbf{y}) + c_{ij\alpha 1}(\mathbf{y}) = 0$. The last equality follows from the symmetry of the elastic constants (see condition *C1*, Sect. 1.2).

Proposition 3.4. *Cellular problem* (3.2.29) *is degenerate. The kernel of problem* (3.2.29) *is formed by the function* **const** $= (C_1, C_2, C_3)$ *and the function*

$$\mathbf{U}(\mathbf{y}) = y_\Gamma s_\gamma \mathbf{e}_\gamma \ . \tag{3.2.31}$$

In (3.2.31), the following notations are used:

$$s_1 = 0,\ s_2 = -1,\ s_3 = 1;$$

$$\Gamma = 2 \text{ if } \gamma = 3\text{; and } \Gamma = 3 \text{ if } \gamma = 2\ .$$

The summation in γ, Γ is assumed (if otherwise not indicated) in accordance with the rule $x_\Gamma y_\gamma = x_3 y_2 + x_2 y_3$.

To verify Proposition 3.4, let us consider the quantity $c_{ijmn} U_{m,ny}$, where $\mathbf{U}(\mathbf{y})$ is given by (3.2.31). It is equal to $c_{ij\gamma\Gamma} s_\gamma = c_{ij23} - c_{ij32} = 0$.

The solutions $\mathbf{X}^{0\alpha}(\mathbf{y})$ (3.2.29) correspond to the macroscopic bending of a beam. The solution $\mathbf{U}(\mathbf{y})$ (3.2.31) corresponds to the macroscopic torsion. From the mathematical point of view, torsion is due to the degeneration of cellular problem (3.2.29) in the local variables in the domain corresponding to the periodicity cell of the beam. No situation like this arises in the cellular problems analyzed above. The cellular problems, considered in connection with the composite solids and plates, are not degenerate (see Chap. 1 for solid composites and Chap. 2 for composite plates).

Allowing for the fact that the function of the argument x_1 plays the role of a parameter in the problems in variables $\mathbf{y}$, and $\mathbf{u}^{(0)}$ depends only on x_1, the solution of problem (3.2.28) and (3.2.27, $k = 1$) can be found in the form,

$$\mathbf{u}^{(1)} = -y_\alpha \mathbf{e}_1 u^{(0)}_{\alpha,1x}(x_1) + \mathbf{X}^{01}(\mathbf{y}) u^{(0)}_{1,1x}(x_1) + y_\Gamma s_\gamma \mathbf{e}_\gamma \phi(x_1) + \mathbf{V}(x_1)\ . \tag{3.2.32}$$

Here, $\mathbf{V}(x_1)$ and $\phi(x_1)$ are arbitrary functions of the argument x_1 (they will be determined below).

From (3.2.32) and (3.2.25, $p = -4$),

$$\sigma^{(-4)}_{ij} = [c_{ijmn}(\mathbf{y}) X^{01}_{m,ny}(\mathbf{y}) + c_{ij11}(\mathbf{y})] u^{(0)}_{1,1x}(x_1)\ . \tag{3.2.33}$$

Averaging (3.2.33) with $ij = 11$, we obtain

$$N_{11} = A u^{(0)}_{1,1x}(x_1)\ , \tag{3.2.34}$$

where

$$A = \langle c_{1111}(\mathbf{y}) + c_{11mn}(\mathbf{y}) X^{01}_{m,ny}(\mathbf{y}) \rangle\ . \tag{3.2.35}$$

Substituting (3.2.34) in homogenized equilibrium equation (3.2.15), we obtain

$$(Au_{1,1x}^{(0)})_{,1x} = 0\,. \tag{3.2.36}$$

Substitution of expansion (3.2.7) in the original boundary conditions (3.2.2) yields the following boundary conditions for $u_1^{(0)}(x_1)$:

$$u_1^{(0)}(-a) = u_1^{(0)}(a) = 0\,. \tag{3.2.37}$$

The solution of problem (3.2.36) and (3.2.37) is $u_1^{(0)}(x_1) = 0$. Then, (3.2.32) takes the form,

$$\mathbf{u}^{(1)} = -y_\alpha \mathbf{e}_1 u_{\alpha,1x}^{(0)}(x_1) + y_\Gamma s_\gamma \mathbf{e}_\gamma \phi(x_1) + \mathbf{V}(x_1)\,. \tag{3.2.38}$$

Substituting (3.2.38) in (3.2.32) gives the following equations:

$$\sigma_{ij}^{(-4)} = 0\,, \tag{3.2.39}$$

$$\sigma_{ij}^{(-3)} = c_{ijmn}(\mathbf{y})u_{m,ny}^{(2)} - c_{ij11}(\mathbf{y})y_\alpha u_{\alpha,1x1x}^{(0)}(x_1) \tag{3.2.40}$$

$$+ c_{ij11}(\mathbf{y})V_{1,1x}(x_1) + c_{ij\gamma 1}(\mathbf{y})s_\gamma y_\Gamma \phi_{,1x}(x_1)\,.$$

Let us examine problem (3.2.23, $p=-3$), (3.2.40), and (3.2.27, $k=2$). In order to solve this problem, we introduce functions $\mathbf{X}^{1\alpha}(\mathbf{y})$ ($\alpha = 2,3$) and $\mathbf{X}^3(\mathbf{y})$ which are determined by solutions of the following cellular problems (corresponding to bending and torsion of beam):

$$\begin{cases} [c_{ijmn}(\mathbf{y})X_{m,ny}^{1\alpha} + c_{ij11}(\mathbf{y})y_\alpha]_{,jy} = 0 \quad \text{in } Y, \\ [c_{ijmn}(\mathbf{y})X_{m,ny}^{1\alpha} + c_{ij11}(\mathbf{y})y_\alpha]n_j = 0 \quad \text{on } S, \\ \mathbf{X}^{1\alpha}(\mathbf{y}) \text{ is periodic in } y_1 \text{ with period } T; \end{cases} \tag{3.2.41}$$

$$\begin{cases} [c_{ijmn}(\mathbf{y})X^3_{m,ny} + c_{ij\gamma 1}(\mathbf{y})s_\gamma y_\Gamma]_{,jy} = 0 & \text{in } Y, \\ [c_{ijmn}(\mathbf{y})X^3_{m,ny} + c_{ij\gamma 1}(\mathbf{y})s_\gamma y_\Gamma]n_j = 0 & \text{on } S, \\ \mathbf{X}^3(\mathbf{y}) \text{ is periodic in } y_1 \text{ with period } T. \end{cases} \tag{3.2.42}$$

The solution of the problem (3.2.23, $p=-3$), (3.2.40), and (3.2.27, $k=2$) can be expressed through the functions $\mathbf{X}^{01}(\mathbf{y})$, $\mathbf{X}^{1\alpha}(\mathbf{y})$, $\mathbf{X}^3(\mathbf{y})$ as follows

$$\begin{aligned} \mathbf{u}^{(2)} &= \mathbf{X}^{1\alpha}(\mathbf{y})u^{(0)}_{\alpha,1x1x}(x_1) - y_\alpha V_{\alpha,1x}(x_1)\mathbf{e}_1 + \mathbf{X}^{01}(\mathbf{y})V_{1,1x}(x_1) \\ &= \mathbf{X}^{1\alpha}(\mathbf{y})u^{(0)}_{\alpha,1x1x}(x_1) - y_\alpha V_{\alpha,1x}(x_1)\mathbf{e}_1 + \mathbf{X}^{01}(\mathbf{y})V_{1,1x}(x_1) \\ &\quad + \mathbf{X}^3(\mathbf{y})\phi_{,1x}(x_1) + \mathbf{U}(x_1) + y_\Gamma s_\gamma \mathbf{e}_\gamma \psi(x_1). \end{aligned} \tag{3.2.43}$$

Writing (3.2.43), we use the equality $c_{ij\gamma\Gamma}s_\gamma = 0$ following from the symmetry of the elastic constants (see condition *C1*, Sect. 1.2). In (3.2.43), $\psi(x_1)$ is a new function (that will not influence the problem).

Substituting (3.2.43) in (3.2.39),

$$\begin{aligned} \sigma^{(-3)}_{ij} &= [c_{ijmn}(\mathbf{y})X^{01}_{m,ny}(\mathbf{y}) - c_{ij11}(\mathbf{y})]V_{1,1x}(x_1) \\ &\quad + [c_{ijmn}(\mathbf{y})X^{1\alpha}_{m,ny}(\mathbf{y}) + c_{ij11}(\mathbf{y})y_\alpha]u^{(0)}_{\alpha,1x1x}(x_1) \\ &\quad + [c_{ijmn}(\mathbf{y})X^3_{m,ny}(\mathbf{y}) + c_{ij\gamma 1}(\mathbf{y})s_\gamma y_\Gamma]\phi_{,1x}(x_1). \end{aligned} \tag{3.2.44}$$

Averaging (3.2.44) with $ij = 11$ over periodicity cell Y, we obtain

$$N^{(-3)}_{11} = AV_{1,1x} + A^1_\alpha u^{(0)}_{\alpha,1x1x} + b\phi_{,1x}. \tag{3.2.45}$$

Multiplying (3.2.44) with $j=1$ by y_β and averaging over periodicity cell Y, we obtain equations, which can be broken down into two groups: equations for bending moments,

$$M_{1\beta}^{(-3)} = \langle \sigma_{11}^{(-3)} y_\beta \rangle ,$$

and equations for torsional moments,

$$M = M_{\Gamma\gamma}^{(-3)} s_\gamma = M_{32}^{(-3)} - M_{23}^{(-3)}$$

$$= \langle \sigma_{\Gamma 1}^{(-3)} \rangle s_\gamma = \langle \sigma_{31}^{(-3)} y_2 - \sigma_{21}^{(-3)} y_3 \rangle .$$

They are the following:

$$M_{1\beta}^{(-3)} = {}^1A_\beta V_{1,1x} + A_{\beta\alpha}^2 u_{\alpha,1x1x}^{(0)} + B_\beta \phi_{,1x} , \tag{3.2.46}$$

$$M = A^1 V_{1,1x} + A_\alpha^2 u_{\alpha,1x1x}^{(0)} + B\phi_{,1x} . \tag{3.2.47}$$

The coefficients in (3.2.45)–(3.2.47) are

$$A = \langle c_{1111}(\mathbf{y}) + c_{11mn}(\mathbf{y}) X_{m,ny}^{01}(\mathbf{y}) \rangle , \tag{3.2.48}$$

$$A_\alpha^1 = \langle c_{1111}(\mathbf{y}) y_\alpha + c_{11mn}(\mathbf{y}) X_{m,ny}^{1\alpha}(\mathbf{y}) \rangle ,$$

$$b = \langle c_{11\gamma 1}(\mathbf{y}) s_\gamma y_\Gamma + c_{11mn}(\mathbf{y}) X_{m,ny}^3(\mathbf{y}) \rangle ,$$

$${}^1A_\beta = \langle [c_{1111}(\mathbf{y}) + c_{11mn}(\mathbf{y}) X_{m,ny}^{11}(\mathbf{y})] y_\beta \rangle ,$$

$$A_{\alpha\beta}^2 = \langle [c_{1111}(\mathbf{y}) y_\alpha + c_{11mn}(\mathbf{y}) X_{m,ny}^{1\alpha}(\mathbf{y})] y_\beta \rangle ,$$

$$B_\beta = \langle [c_{11\gamma 1}(\mathbf{y}) s_\gamma y_\Gamma + c_{11mn}(\mathbf{y}) X_{m,ny}^3(\mathbf{y})] y_\beta \rangle ,$$

$$B = \langle [c_{B1\gamma 1}(\mathbf{y}) y_\Gamma s_\gamma + c_{B1mn}(\mathbf{y}) X_{m,ny}^3(\mathbf{y})] y_\beta \rangle s_\beta ,$$

$$A^1 = \langle [c_{3111}(\mathbf{y}) + c_{31mn}(\mathbf{y}) X_{m,ny}^{01}(\mathbf{y})] y_2 \rangle$$

$$- \langle [c_{2111}(\mathbf{y}) + c_{21mn}(\mathbf{y}) X_{m,ny}^{01}(\mathbf{y})] y_3 \rangle ,$$

$$A_\alpha^2 = \langle [c_{3111}(\mathbf{y})y_\alpha + c_{31mn}(\mathbf{y})X_{m,ny}^{1\alpha}(\mathbf{y})]y_2\rangle$$

$$-\langle [c_{2111}(\mathbf{y})y_\alpha + c_{21mn}(\mathbf{y})X_{m,ny}^{1\alpha}(\mathbf{y})]y_3\rangle .$$

The equations (3.2.45)–(3.2.47) obtained are asymptotically exact governing equations of the beam considered as a 1-D structure. The coefficients in the right-hand parts of (3.2.45)–(3.2.47) are the homogenized stiffnesses of the beam. They are expressed through solutions of cellular problems (3.2.29), (3.2.41), and (3.2.42). Note that the homogenized governing equations derived coincide with the classical ones in form, but the homogenized stiffnesses are calculated in anther way.

Let us denote the right-hand parts of the governing equations (3.2.45)–(3.2.47) by

$$N(V_{1,1x}, u_{\alpha,1x1x}^{(0)}, \phi_{,1x}) = AV_{1,1x} + A_\alpha^1 u_{\alpha,1x1x}^{(0)} + b\phi_{,1x} ,$$

$$M_\beta(V_{1,1x}, u_{\alpha,1x1x}^{(0)}, \phi_{,1x}) = {}^1A_\beta V_{1,1x} + A_{\beta\alpha}^2 u_{\alpha,1x1x}^{(0)} + B_\beta \phi_{,1x} ,$$

$$M(V_{1,1x}, u_{\alpha,1x1x}^{(0)}, \phi_{,1x}) = A^1 V_{1,1x} + A_\alpha^2 u_{\alpha,1x1x}^{(0)} + B\phi_{,1x} ,$$

respectively. These notations will be used in the next sections.

Excluding the shear forces

We collect the equilibrium equations (3.2.15, $p=-3$), (3.2.16), (3.2.20), and (3.2.21):

$$N_{11,1x}^{(-3)} + \langle f_1\rangle + \langle g_1\rangle_S = 0 , \tag{3.2.49}$$

$$N_{\beta1,1x}^{(-2)} + \langle f_\beta\rangle + \langle g_\beta\rangle_S = 0 , \tag{3.2.50}$$

$$M_{1\beta,1x}^{(-3)} - N_{1\beta}^{(-2)} + \langle f_1 y_\beta\rangle + \langle g_1 y_\beta\rangle_S = 0 , \tag{3.2.51}$$

$$M_{,1x} + (N_{32}^{(-2)} - N_{23}^{(-2)}) = 0 . \tag{3.2.52}$$

In the absence of initial stresses, we have the equality $\sigma_{ij}^{(-2)} = \sigma_{ji}^{(-2)}$. Thus $N_{ij}^{(-2)} = N_{ji}^{(-2)}$, and we can exclude the quantities $N_{1\beta}^{(-2)}$ and $N_{\beta 1}^{(-2)}$ from (3.2.50) and (3.2.51) and quantity $N_{32}^{(-2)} - N_{23}^{(-2)}$ from (3.2.52) using this symmetry only with respect to indexes i and j. In our case, $N_{ij}^{(-2)}$ does not have symmetry with respect to indices i and j. Then, we need to examine $N_{ij}^{(-2)}$ in detail, to obtain some additional information on them.

Let us insert $\mathbf{u}^{(1)}$ into (3.2.26) in accordance with (3.2.38). Then, we obtain

$$\sigma_{ij}^{(-2)} = c_{ijm1}(\mathbf{y})u_{m,1x}^{(2)} + c_{ijmn}(\mathbf{y})u_{m,ny}^{(3)} \tag{3.2.53}$$

$$+ b_{ij\alpha 1}(x_1, \mathbf{y})u_{\alpha,1x}^{(0)}(x_1) + b_{ij\gamma\Gamma}(x_1, \mathbf{y})s_\gamma \phi(x_1).$$

The first and second terms on the right-hand side of (3.2.53) are symmetrical with respect to i and j by virtue of the symmetry of the elastic constants (see condition *C1 a*). Then, the relation below follows from (3.2.53)

$$\sigma_{ij}^{(-2)} - \sigma_{ji}^{(-2)} = b_{ij\alpha 1}(x_1, \mathbf{y})u_{\alpha,1x}^{(0)}(x_1) - b_{ji\alpha 1}(x_1, \mathbf{y})u_{\alpha,1x}^{(0)}(x_1) \tag{3.2.54}$$

$$+ b_{ij\gamma\Gamma}(x_1, \mathbf{y})s_\gamma \phi(x_1) - b_{ji\gamma\Gamma}(x_1, \mathbf{y})s_\gamma \phi(x_1).$$

Averaging the equality (3.2.54) over periodicity cell Y, we obtain

$$N_{ij}^{(-2)} - N_{ji}^{(-2)} = \langle b_{ij\alpha 1}(x_1, \mathbf{y})\rangle u_{\alpha,1x}^{(0)}(x_1) - \langle b_{ji\alpha 1}(x_1, \mathbf{y})\rangle u_{\alpha,1x}^{(0)}(x_1) \tag{3.2.55}$$

$$+ \langle b_{ij\gamma\Gamma}(x_1, \mathbf{y})\rangle s_\gamma \phi(x_1) - \langle b_{ji\gamma\Gamma}(x_1, \mathbf{y})\rangle s_\gamma \phi(x_1).$$

To continue our investigation, we need some facts about the average values of the initial stresses.

Proposition 3.5. *Let initial stresses* $\overset{*}{\sigma}_{ij}$, *which can be represented as* $\overset{*}{\sigma}_{ij} = \varepsilon^{-3}\sigma_{ij}^{*(-3)} + \varepsilon^{-2}\sigma_{ij}^{*(-2)} + \ldots$, *satisfy the equilibrium equations,*

$$\frac{\partial \overset{*}{\sigma}_{ij}}{\partial x_j} + \varepsilon^a f_i(x_1, \mathbf{y}) = 0 \quad \text{in } G_\varepsilon, \tag{3.2.56}$$

$$\overset{*}{\sigma}_{ij} n_j = \varepsilon^b g_i(x_1, \mathbf{y}) \text{ on } S_\varepsilon .$$

Then,

1) $\langle b^{*(-3)}_{ijmn}(x_1, \mathbf{y}) Z_{,ny} \rangle = 0$ *if* $a < -4$, $b < -3$;

2) $\langle b^{*(-2)}_{ijmn}(x_1, \mathbf{y}) Z_{,ny} \rangle = 0$ *if* $\sigma^{*(-3)}_{ij} = 0$ *and* $a < -3$, $b < -2$,

for every differentiable function $Z(\mathbf{y})$ *periodic in* y_1 *with period* T.

Taking into account the definition of $b^{*(-3)}_{ijmn}$ (3.2.4), one must prove that $\langle \sigma^{*(-3)}_{jn} Z_{,ny} \rangle = 0$. Substituting the expression $\overset{*}{\sigma}_{ij} = \varepsilon^{-3} \sigma^{*(-3)}_{ij} + \varepsilon^{-2} \sigma^{*(-2)}_{ij} + \ldots$ for the initial stresses in (3.2.56), with allowance for (3.2.9), we obtain

$$\overset{*}{\sigma}_{ij,jy} + F_i(x_1, \mathbf{y}) = 0 \quad \text{in } Y \ (F_i = 0 \text{ if } a \neq -4,\ F_i = f_i \text{ if } a = -4), \qquad (3.2.57)$$

$$\overset{*}{\sigma}_{ij} n_j = G_i(x_1, \mathbf{y}) \quad \text{on } S \ \ (G_i = 0 \text{ if } a \neq -3,\ G_i = g_i \text{ if } a = -3),$$

$$\overset{*}{\sigma}_{ij}(x_1, \mathbf{y}) \text{ is periodic in } y_1 \text{ with period } T.$$

Let us consider the quantity $\langle \overset{*}{\sigma}_{jn}(x_1, \mathbf{y}) Z_{,ny}(\mathbf{y}) \rangle$. Taking into account the definition of the average value over periodicity cell Y and integrating by parts, one can find that this quantity is equal to

$$-\int_Y \overset{*}{\sigma}_{jn,ny} Z(\mathbf{y}) d\mathbf{y} + \int_S \overset{*}{\sigma}_{jn} n_n Z(\mathbf{y}) d\mathbf{y} + \int_\omega \overset{*}{\sigma}_{j1} n_1 Z(\mathbf{y}) d\mathbf{y} =$$

$$\int_Y F_j Z(\mathbf{y}) d\mathbf{y} + \int_S g_j Z(\mathbf{y}) d\mathbf{y}.$$

Here, $\omega = \omega_+ \cup \omega_-$ mean the opposite faces of periodicity cell Y normal to the Oy_1 axis (see Fig. 3.2). The integrals over Y and S are equal to zero as a consequence of (3.2.57) (if $a \neq -4$, $b \neq -3$). The integral over ω is equal to zero because of the periodicity of functions $\sigma^{*(-3)}_{jn}$ and Y and the antiperiodicity of the vector normal in y_1 (see Fig. 3.2). That proves statement 1 of Proposition 3.5. Analogously, one can prove statement 2 of Proposition 3.5.

Proposition 3.6. *Under the conditions of Proposition 3.5,* $\langle \sigma^{*(-3)}_{i\alpha} \rangle = 0$, *and* $\langle \sigma^{*(-3)}_{1\alpha} y_\alpha \rangle = 0$, *and* $\langle \sigma^{*(-2)}_{i\alpha} \rangle = 0$, *when* $\langle \sigma^{*(-3)}_{ij} \rangle = 0$.

Proposition 3.6 is a consequence of Proposition 3.5. To prove the first and the third equations, one set $Z = y_\alpha$, and $Z = y_\alpha^2$ to prove the second one.

Now, we can exclude shear forces $N_{\beta 1}^{(-2)}$ and $N_{1\beta}^{(-2)}$ from the equilibrium equations.

Bending

Consider the following equilibrium equations (3.2.50) and (3.2.51). To exclude $N_{\beta 1}^{(-2)}$ and $N_{1\beta}^{(-2)}$, apply the relations, which follow from (3.2.55) and the definition of $b_{ijmn}^{(-2)}$ (3.2.4):

$$N_{1\beta}^{(-2)} - N_{\beta 1}^{(-2)} = K_\beta, \tag{3.2.58}$$

where $K_\beta(x_1)$ is defined by the formula

$$K_\beta = \langle b_{\beta 1\alpha 1}(x_1,\mathbf{y})\rangle u_{\alpha,1x}^{(0)}(x_1) - \langle b_{1\beta\alpha 1}(x_1,\mathbf{y})\rangle u_{\alpha,1x}^{(0)}(x_1)$$

$$+\langle b_{\beta 1\gamma\Gamma}(x_1,\mathbf{y})\rangle s_\gamma \phi(x_1) - \langle b_{1\beta\gamma\Gamma}(x_1,\mathbf{y})\rangle s_\gamma \phi(x_1) \tag{3.2.59}$$

$$= \langle \overset{*}{\sigma}_{11}(x_1,\mathbf{y})\rangle \delta_{\beta\alpha} u_{\alpha,1x}^{(0)}(x_1) + \langle \overset{*}{\sigma}_{1\Gamma}(x_1,\mathbf{y})\rangle \delta_{\beta\gamma} s_\gamma \phi(x_1).$$

Differentiating (3.2.50) and using (3.2.51) and (3.2.58), we obtain

$$(M_{1\beta,1x}^{(-3)} + \langle f_1 y_\beta\rangle + \langle g_1 y_\beta\rangle_S)_{,1x} = K_{\beta,1x} - \langle f_\beta\rangle - \langle g_\beta\rangle_S. \tag{3.2.60}$$

Under the conditions of Proposition 3.5 ($a < -3$, $b < -2$)

$$K_\beta = \langle \overset{*}{\sigma}_{11}(x_1,\mathbf{y})\rangle u_{\beta,1x}^{(0)}(x_1),$$

then (3.2.60) takes the classical form ,

$$M_{1\beta,1x1x}^{(-3)} + \langle f_\beta\rangle + \langle g_\beta\rangle_S + (\langle f_1 y_\beta\rangle + \langle g_1 y_\beta\rangle_S)_{,1x} \tag{3.2.61}$$

$$= [\overset{*}{N}_{11} u_{\beta,1x}^{(0)}(x_1)]_{,1x},$$

where $\overset{*}{N}_{11}(x_1) = \langle \overset{*}{\sigma}_{11}(x_1,\mathbf{y})\rangle$ is the initial axial force.

Torsion

Here, as above, we encounter a situation connected with the asymmetry of $N_{ij}^{(-2)}$. Due to the asymmetry, we have an unknown (and non-zero) term $N_{32}^{(-2)} - N_{23}^{(-2)}$ in (3.2.52). In accordance with (3.2.55) and the definition of $b_{ijmn}^{(-2)}$ (3.2.4), we can write

$$N_{23}^{(-2)} = N_{32}^{(-2)} + K\,, \tag{3.2.62}$$

where

$$K = N_{32}^{(-2)} - N_{23}^{(-2)} = \langle b_{32\alpha 1}(x_1, \mathbf{y})\rangle u_{\alpha,1x}^{(0)}(x_1) - \langle b_{23\alpha 1}(x_1, \mathbf{y})\rangle u_{\alpha,1x}^{(0)}(x_1) \tag{3.2.63}$$

$$+\langle b_{32\gamma\Gamma}(x_1, \mathbf{y})\rangle s_\gamma \phi(x_1) - \langle b_{23\gamma\Gamma}(x_1, \mathbf{y})\rangle s_\gamma \phi(x_1)$$

$$= -\langle \overset{*}{\sigma}_{31}(x_1, \mathbf{y})\rangle \delta_{2\alpha} u_{\alpha,1x}^{(0)}(x_1) - \langle \overset{*}{\sigma}_{3\Gamma}(x_1, \mathbf{y})\rangle \delta_{2\gamma} s_\gamma \phi(x_1)\,.$$

Then, (3.2.52) can be rewritten in the form,

$$M_{1,1x} - \mathrm{K} = 0 \tag{3.2.64}$$

Under the conditions of Proposition 3.5 ($a \neq -3$, $b \neq -2$), $K = 0$. This means, that the initial stresses of the order ε^{-2} do not influence the torsion of the beam.

The limit problem

With regard to the homogenized governing equations (3.2.45)–(3.2.47), we can write the homogenized equations of equilibrium (3.2.49), (3.2.61), (3.2.64) in the form,

$$(AV_{1,1x} + A_\alpha^1 u_{\alpha,1}^{(0)} + b\phi_{,1x})_{,1x} + \langle f_1\rangle + \langle g_1\rangle_S = 0\,, \tag{3.2.65}$$

$$({}^1A_\beta V_{1,1x} + A_{\beta\alpha}^2 u_{\alpha,1x1x}^{(0)} + B_\beta \phi_{,1x})_{,1x1x} \tag{3.2.66}$$

$$+\langle f_\beta\rangle + \langle g_\beta\rangle_S + (\langle f_1 y_\beta\rangle + \langle g_1 y_\beta\rangle_S)_{,1x} = (\overset{*}{N}_{11} u_{\alpha,1x}^{(0)})_{,1x}\,,$$

$$(A^1 V_{1,1x} + A_\alpha^2 u_{\alpha,1x1x}^{(0)} + B\phi_{,1x})_{,1x} = 0\,. \tag{3.2.67}$$

Here, $N_{11}^{*} = \langle \sigma_{11}^{*} \rangle$.

The boundary conditions are the following:

$$V_1(-a) = V_1(a) = 0, \tag{3.2.68}$$

$$u_{\alpha}^{(0)}(-a) = u_{\alpha}^{(0)}(a) = u_{\alpha,1x}^{(0)}(-a) = u_{\alpha,1x}^{(0)}(a) = 0,$$

$$\phi(-a) = \phi(a) = 0 .$$

Problem (3.2.65)–(3.2.67) with boundary conditions (3.2.68) is a 1-D model of a beam with respect to the functions $u_{\alpha}^{(0)}(x_1)\,(\alpha = 2,3)$, $V_1(x_1)$, and $\phi(x_1)$. Model (3.2.65)–(3.2.68) coincides qualitatively with the classical equations.

The case of uniform homogenized initial stresses

Let the initial stresses depend only on $\mathbf{x}/\varepsilon$ (do not depend on x_1). Then, the average value over periodicity cell Y does not depend on x_1 (see, e.g., Kalamkarov and Kolpakov, 1997), $N_{11}^{*(-2)}$ is a constant, and (3.2.66) can be written as

$$({}^1A_{\beta}V_{1,1x} + A_{\beta\alpha}^{2}u_{\alpha,1x1x}^{(0)} + B_{\beta}\phi_{,1x})_{,1x1x} \tag{3.2.69}$$

$$+ \langle f_{\beta} \rangle + \langle g_{\beta} \rangle_S + (\langle f_1 y_{\beta} \rangle + \langle g_1 y_{\beta} \rangle_S)_{,1x} = N_{11}^{*}u_{\alpha,1x1x}^{(0)} .$$

Cylindrical beams. Comparison with the classical case

As demonstrated by Kolpakov (1994) for a classical cylindrical beam made of isotropic material, the stiffnesses A, A_{α}^{1}, b, ${}^1A_{\alpha}$, $A_{\alpha\beta}^{2}$, B_{β}, B coincide with obtained using the plane section hypothesis. Also in this case, $\langle f_1 y_{\beta} \rangle = \langle g_1 y_{\beta} \rangle_S = 0$. Then, (3.2.65)–(3.2.67) coincide with the classical equations both qualitatively and quantitatively.

3.3 Stressed Beams (Transition from 3-D to 1-D Model, Moments of Initial Stresses)

We consider in this section the following case: $\{\sigma_{ij}^{*(-3)} \neq 0, \sigma_{ij}^{*(-2)} = 0\}$, and $\sigma_{ij}^{*(-3)} \neq 0$ satisfies the condition

$$\langle \sigma_{11}^{*(-3)} \rangle = 0 \tag{3.3.1}$$

- zero axial force corresponding to local stresses $\sigma_{11}^{*(-3)}$.

Furthermore, in this section, σ_{ij}^{*} means $\sigma_{ij}^{*(-3)}$, and b_{ijkl} means $b_{ijkl}^{(-3)}$.

As demonstrated in Sect. 3.2, the homogenized equations of equilibrium do not depend on local governing equation (3.2.5). Then, in the case under consideration, we must concentrate on the problem of homogenized governing equations.

Homogenized governing equations

We use asymptotic expansions (3.2.7) and (3.2.8). Substituting (3.2.7) and (3.2.8) in local governing equation (3.2.5), with allowance for (3.2.9), we obtain in the case under consideration,

$$\sum_{p=-4}^{\infty} \varepsilon^{p} \sigma_{ij}^{(p)} = \sum_{k=0}^{\infty} \varepsilon^{k} (\varepsilon^{-4} c_{ijmn}(\mathbf{y}) + \varepsilon^{-3} b_{ijmn}(x_1, \mathbf{y}))(u_{m,1x}^{(k)} \delta_{n1} + \varepsilon^{-1} u_{m,ny}^{(k)}). \tag{3.3.2}$$

Equating the terms with identical powers of ε in (3.3.2), we obtain

$$\sigma_{ij}^{(-4)} = c_{ijm1}(\mathbf{y}) u_{m,1x}^{(0)} + c_{ijmn}(\mathbf{y}) u_{m,ny}^{(1)}, \tag{3.3.3}$$

$$\sigma_{ij}^{(-3)} = c_{ijm1}(\mathbf{y}) u_{m,1x}^{(1)} + b_{ijm1}(x_1, \mathbf{y}) u_{m,1x}^{(0)} + c_{ijmn}(\mathbf{y}) u_{m,ny}^{(2)} + b_{ijmn}(x_1, \mathbf{y}) u_{m,ny}^{(1)}, \tag{3.3.4}$$

$$\sigma_{ij}^{(-2)} = c_{ijm1}(\mathbf{y}) u_{m,1x}^{(2)} + b_{ijm1}(x_1, \mathbf{y}) u_{m,1x}^{(1)} + c_{ijmn}(\mathbf{y}) u_{m,ny}^{(3)} + b_{ijmn}(x_1, \mathbf{y}) u_{m,ny}^{(2)}. \tag{3.3.5}$$

Let us consider problem (3.2.23, $p=-4$) with $\sigma_{ij}^{(-4)}$ given by (3.3.3) and condition (3.2.27, $k=1$). It coincides with problem (3.2.23, $p=-4$), (3.2.25, $p=-4$), (3.2.27, $k=1$) studied in Sect. 3.2. Then, its solution $\mathbf{u}^{(1)}$ is given by (3.2.38).

Substituting $\mathbf{u}^{(1)}$ (3.2.38) in (3.3.3) and (3.3.4) gives the following equations:

$$\sigma_{ij}^{(-4)} = 0, \tag{3.3.6}$$

$$\sigma_{ij}^{(-3)} = c_{ijmn}(\mathbf{y})u_{m,ny}^{(2)} - c_{ij11}(\mathbf{y})y_\alpha u_{\alpha,1x1x}^{(0)}(x_1) \tag{3.3.7}$$

$$+ c_{ij11}(\mathbf{y})V_{1,1x}(x_1) + c_{ij\gamma 1}(\mathbf{y})s_\gamma y_\Gamma \phi_{,1x}(x_1)$$

$$+ (b_{ij\alpha 1}(x_1,\mathbf{y}) - b_{ij1\alpha}(x_1,\mathbf{y}))u_{m,1x}^{(0)}(x_1) + b_{ij\gamma\Gamma}(x_1,\mathbf{y})s_\gamma \phi(x_1).$$

Under the conditions of Propositions 3.6, see (3.2.57),

$$b_{ijmn,jy} = \overset{*}{\sigma}_{nj,jy}\delta_{im} = 0 \text{ in } Y, \text{ and } b_{ijmn}n_j = \overset{*}{\sigma}_{nj}\delta_{im}n_j = 0 \quad \text{on } S.$$

Then, the terms containing b_{ijmn} have no influence on the solution, and, solving the problem (3.2.23, $p=-3$) with $\sigma_{ij}^{(-3)}$ given by (3.3.7), one can set

$$\sigma_{ij}^{(-3)} = c_{ijmn}(\mathbf{y})u_{m,ny}^{(2)} + c_{ij11}(\mathbf{y})y_\alpha u_{\alpha,1x1x}^{(0)}(x_1) \tag{3.3.8}$$

$$+ c_{ij11}(\mathbf{y})V_{1,1x}(x_1) + c_{ij\gamma 1}(\mathbf{y})s_\gamma y_\Gamma \phi_{,1x}(x_1).$$

This expression coincides with (3.2.40). Then the solution of the problem under consideration is given by (3.2.43). Substituting $\mathbf{u}^{(2)}$ (3.2.43) in (3.3.4), we obtain

$$\sigma_{ij}^{(-3)} = (c_{ijmn}(\mathbf{y})X_{m,ny}^{01} + c_{ij11}(\mathbf{y}))V_{1,1x}(x_1) \tag{3.3.9}$$

$$+ (c_{ijmn}(\mathbf{y})X_{m,ny}^{1\alpha} + c_{ij11}(\mathbf{y})y_\alpha)u_{\alpha,1x1x}^{(0)}(x_1)$$

$$+ (c_{ijmn}(\mathbf{y})X_{m,ny}^{3} + c_{ij\gamma 1}(\mathbf{y})s_\gamma y_\Gamma)\phi_{,1x}(x_1)$$

$$+(b_{ij\alpha 1}(x_1,\mathbf{y})-b_{ij1\alpha}(x_1,\mathbf{y}))u^{(0)}_{\alpha,1x}(x_1)$$

$$+b_{ij\gamma\Gamma}(x_1,\mathbf{y})s_{,\gamma}\psi(x_1).$$

Averaging (3.3.9) with $ij=11$ over periodicity cell Y and taking into account the definition of b_{ijmn} (3.2.4) and Proposition 3.6, we obtain

$$N^{(-3)}_{11}=N(V_{1,1x},\ u^{(0)}_{\alpha,1x1x},\ \phi_{,1x}), \tag{3.3.10}$$

where $N(V_{1,1x},\ u^{(0)}_{\alpha,1x1x},\ \phi_{,1x})$ is determined in Sect. 3.2; see (3.2.45).

Multiplying (3.3.9) with $j=1$ by y_β and averaging over periodicity cell Y, we obtain, taking into account the definition of b_{ijmn} (3.2.4), the definition of the bending and torsional moments, condition (3.3.1), and Propositions 3.6 and 3.7,

$$M^{(-3)}_{1\beta}=M_\beta(V_{1,1x},\ u^{(0)}_{\alpha,1x1x},\ \phi_{,1x})+\mathbf{C}_{\alpha\beta}u^{(0)}_{\alpha,1x}, \tag{3.3.11}$$

$$M=M(V_{1,1x},\ u^{(0)}_{\alpha,1x1x},\ \phi_{,1x})+\mathbf{c}_\alpha u^{(0)}_{\alpha,1x}. \tag{3.3.12}$$

Here $M_\beta(V_{1,1x},\ u^{(0)}_{\alpha,1x1x},\ \phi_{,1x})$ and $M(V_{1,1x},\ u^{(0)}_{\alpha,1x1x},\ \phi_{,1x})$ are determined in Sect. 3.2, (3.2.46) and (3.2.47).

The coefficients on the right-hand side of (3.3.11) and (3.3.12) are the following:

$$\mathbf{C}_{\beta\alpha}=\langle[b_{11\alpha 1}(x_1,\mathbf{y})-b_{111\alpha}(x_1,\mathbf{y})]y_\beta\rangle=-\langle\overset{*}{\sigma}_{1\alpha}(x_1,\mathbf{y})y_\beta\rangle, \tag{3.3.13}$$

$$\mathbf{c}_\alpha=\langle[b_{31\alpha 1}(x_1,\mathbf{y})-b_{311\alpha}(x_1,\mathbf{y})]y_2-[b_{21\alpha 1}(x_1,\mathbf{y})-b_{211\alpha}(x_1,\mathbf{y})]y_3\rangle$$

$$=\langle\overset{*}{\sigma}_{11}(x_1,\mathbf{y})y_2\rangle\delta_{3\alpha}-\langle\overset{*}{\sigma}_{11}(x_1,\mathbf{y})y_3\rangle\delta_{2\alpha}(x_1,\mathbf{y}).$$

Deriving (3.3.11), (3.3.12), and (3.3.13), we take into account the definition of b_{ijkl} (3.2.4) and use the equality

$$\langle b_{31\gamma\Gamma}(x_1,\mathbf{y})s_{,\gamma}y_2\rangle-\langle b_{21\gamma\Gamma}(x_1,\mathbf{y})s_{,\gamma}y_3\rangle$$

$$= -\langle \sigma^*_{12}(x_1, \mathbf{y}) y_2 \rangle - \langle \sigma^*_{31}(x_1, \mathbf{y}) y_3 \rangle = 0 \, ,$$

following from Proposition 3.6.

Excluding the shear forces

Let us exclude the quantities $N^{(-2)}_{ij}$ from equilibrium equations (3.2.50) –(3.2.52). To do this, we examine $N^{(-2)}_{ij}$. In the case under consideration [see (3.3.5)], $\sigma^{(-2)}_{ij}$ is the sum of the terms symmetrical with respect to i and j [the first and third terms in (3.3.5)] and the terms asymmetrical with respect to i and j [the second and fourth terms in (3.3.5)]. One can find that the relation below holds:

$$N^{(-2)}_{ij} = N^{(-2)}_{ji} + \langle [b_{ijm1}(x_1, \mathbf{y}) - b_{jim1}(x_1, \mathbf{y})] u^{(1)}_{m,1x} \rangle \qquad (3.3.14)$$

$$+ \langle [b_{ijmn}(x_1, \mathbf{y}) - b_{jimn}(x_1, \mathbf{y})] u^{(2)}_{m,nx} \rangle \, .$$

Proposition 3.7. *Under the conditions of Proposition 3.6,* $\langle (b_{ijmn} - b_{jimn}) u^{(2)}_{m,nx} \rangle = 0$.

To prove the proposition, it is enough to put $Z = u^{(2)}_m$ into Proposition 3.5.

Substituting (3.2.38) in (3.3.14) and taking into account Proposition 3.7, we can rewrite (3.3.14) as

$$N^{(-2)}_{ij} = N^{(-2)}_{ji} + \langle [b_{ij11}(x_1, \mathbf{y}) - b_{ji11}(x_1, \mathbf{y})] u^{(1)}_{m,1x} \rangle \qquad (3.3.15)$$

$$+ \langle [b_{ij\gamma 1}(x_1, \mathbf{y}) - b_{ji\gamma 1}(x_1, \mathbf{y})] y_\Gamma \rangle s_\gamma \phi_{,1x}(x_1)$$

$$+ \langle b_{ijm1}(x_1, \mathbf{y}) - b_{jim1}(x_1, \mathbf{y}) \rangle V_{m,1x}(x_1) \, .$$

In accordance with condition (3.3.1), the last term on the right-hand side of (3.3.15) is equal to zero.

Applying (3.3.15), we can exclude the quantities $N^{(-2)}_{ij}$ from equilibrium equations (3.2.50), (3.2.51).

Bending

Taking into account (3.3.11), (3.3.15), we can exclude $N_{1\beta}^{(-2)}$ and $N_{\beta 1}^{(-2)}$ from (3.2.50), (3.2.51) and obtain the following equation:

$$M_{1\beta}^{(-3)} = M_\beta(V_{1,1x}, u_{\alpha,1x1x}^{(0)}, \phi_{,1x}) + \mathbf{C}_{\alpha\beta} u_{\alpha,1x}^{(0)} = K_{\beta,1x}^m . \qquad (3.3.16)$$

Here, we denote

$$K_\beta^m = N_{1\beta}^{(-2)} - N_{\beta 1}^{(-2)} = k_{\beta\alpha} u_{\alpha,1x1x}^{(0)} + k_\beta \phi_{,1x} , \qquad (3.3.17)$$

where the following notations are used:

$$k_{\beta\alpha} = \langle (b_{1\beta 11}(x_1, \mathbf{y}) - b_{\beta 111}(x_1, \mathbf{y})) y_\alpha \rangle = \langle \sigma_{\beta 1}(x_1, \mathbf{y}) y_\alpha \rangle , \qquad (3.3.18)$$

$$k_\beta = \langle (b_{1\beta\gamma 1}(x_1, \mathbf{y}) - b_{\beta 1\gamma 1}(x_1, \mathbf{y})) y_\Gamma \rangle s_\gamma = \langle \sigma_{11}(x_1, \mathbf{y}) y_\beta \rangle s_\mathrm{B}$$

(no sum in β, B here),

Here, $\mathrm{B} = 3$ if $\beta = 2$, and $\mathrm{B} = 2$ if $\beta = 3$; $M_\beta(V_{1,1x}, u_{\alpha,1x1x}^{(0)}, \phi_{,1x})$ is determined in Sect. 3.2.

Torsion

We can derive the following equation from (3.2.52), (3.3.12), and (3.3.15):

$$[M(V_{1,1x}, u_{\alpha,1x1x}^{(0)}, \phi_{,1x}) + \mathbf{c}_\alpha u_{\alpha,1x}^{(0)}]_{,1x} = K^m . \qquad (3.3.19)$$

Here,

$$K^m = N_{32}^{(-2)} - N_{23}^{(-2)} = k\phi_{,1x} , \qquad (3.3.20)$$

where

$$k = \langle [b_{32\gamma 1}(x_1, \mathbf{y}) - b_{23\gamma 1}(x_1, \mathbf{y})] y_\Gamma s_\gamma \rangle \qquad (3.3.21)$$

$$= \langle \sigma_{21}(x_1, \mathbf{y}) y_2 + \sigma_{31}(x_1, \mathbf{y}) y_3 \rangle ,$$

$M(V_{1,1x}, u_{\alpha,1x1x}^{(0)}, \phi_{,1x})$ is determined in Sect. 3.2.

Note that all the coefficients determined by (3.3.13), (3.3.17), and (3.3.20) could be written as the following combinations of moments of initial stresses $M^{*(-3)}_{i\alpha} = \langle \sigma^{*(-3)}_{i\alpha} y_\alpha \rangle$:

$$\mathbf{C}_{\beta\alpha} = M^{*(-3)}_{\beta\alpha}, \quad \mathbf{c}_\alpha = M^{*(-3)}_{12}\delta_{3\alpha} - M^{*(-3)}_{13}\delta_{2\alpha}, \tag{3.3.22}$$

$$k_{\beta\alpha} = M^{*(-3)}_{\beta\alpha}, \quad k_\beta = -M^{*(-3)}_{1\beta} s_{\mathrm{B}} \quad \text{(no sum in } \beta, \mathrm{B} \text{ here)},$$

$$k = M^{*(-3)}_{22} + M^{*(-3)}_{33}.$$

By virtue of the symmetry of initial stresses σ^*_{ij} with respect to i and j and Proposition 3.6, $M^{*(-3)}_{\beta\beta} = 0$. Then, $C_{\beta\beta} = 0$, $J = 0$, $k = 0$, and the nonzero coefficients in (3.3.22) are the following:

$$\mathbf{C}_{\beta\mathrm{B}} = M^{*(-3)}_{\beta\mathrm{B}}, \quad \mathbf{c}_\alpha = M^{*(-3)}_{12}\delta_{3\alpha} - M^{*(-3)}_{13}\delta_{2\alpha}, \tag{3.3.23}$$

$$k_{\beta\mathrm{B}} = -M^{*(-3)}_{\beta\mathrm{B}}, \quad k_\beta = M^{*(-3)}_{1\beta} s_{\mathrm{B}} \quad \text{(no sum in } \beta, \mathrm{B} \text{ here)}.$$

Index B was determined in the previous section.

The limit 1-D problem

Let us write the 1-D equations of equilibrium and boundary conditions obtained. The equations of the limit model can be written in the following form:

$$(AV_{1,1x} + A^1_\alpha u^{(0)}_{\alpha,1} + b\phi_{,1x})_{,1x} + \langle f_1 \rangle + \langle g_1 \rangle_S = 0. \tag{3.3.24}$$

$$({}^1A_\beta V_{1,1x} + A^2_{\beta\alpha} u^{(0)}_{\alpha,1x1x} + B_\beta \phi_{,1x})_{,1x1x} \tag{3.3.25}$$

$$+\langle f_\beta \rangle + \langle g_\beta \rangle_S + (\langle f_1 y_\beta \rangle + \langle g_1 y_\beta \rangle_S)_{,1x}$$

$$= (-\mathbf{C}_{\beta\mathrm{B}} u^{(0)}_{\mathrm{B},1x})_{,1x1x} + (k_{\beta\mathrm{B}} u^{(0)}_{\mathrm{B},1x1x} + k_\beta \phi_{,1x})_{,1x},$$

$$(A^1 V_{1,1x} + A^2_\alpha u^{(0)}_{\alpha,1x1x} + B\phi_{,1x})_{,1x} = (\mathbf{c}_\alpha u^{(0)}_{\alpha,1x})_{,1x}, \tag{3.3.26}$$

$\mathbf{C}_{\beta \mathrm{B}}$, $\mathbf{C}_\alpha$, $k_{\beta \mathrm{B}}$, k_β are determined by (3.3.23); no sum in β, B in (3.3.25).

The boundary conditions are

$$V_1(-a) = V_1(a) = 0, \tag{3.3.27}$$

$$u_\alpha^{(0)}(-a) = u_\alpha^{(0)}(a) = u_{\alpha,1x}^{(0)}(-a) = u_{\alpha,1x}^{(0)}(a) = 0,$$

$$\phi(-a) = \phi(a) = 0\,.$$

Model (3.3.24)–(3.3.26) coincides qualitatively with the classical equations.

The case of uniform homogenized initial stresses

Let the initial stresses depend only on $\mathbf{x}/\varepsilon$ (do not depend on x_1). Then, the average value over periodicity cell Y does not depend on x_1 (see, e.g., Kalamkarov and Kolpakov (1997)) and $\mathbf{C}_{\beta \mathrm{B}}$, $\mathbf{c}_\alpha$, $k_{\beta \mathrm{B}}$, k_β are constants. Then the equations (3.3.25) and (3.2.26) can be written as

$$({}^1\!A_\beta V_{1,1x} + A^2_{\beta\alpha} u^{(0)}_{\alpha,1x1x} + B_\beta \phi_{,1x})_{,1x1x} \tag{3.3.28}$$

$$+\langle f_\beta\rangle + \langle g_\beta\rangle_S + (\langle f_1 y_\beta\rangle + \langle g_1 y_\beta\rangle_S)_{,1x}$$

$$= (-\mathbf{C}_{\beta \mathrm{B}} + k_{\beta \mathrm{B}}) u^{(0)}_{\mathrm{B},1x1x1x} + k_\beta \phi_{,1x1x}\,,$$

$$(A^1 V_{1,1x} + A^2_\alpha u^{(0)}_{\alpha,1x1x} + B\phi_{,1x})_{,1x} = \mathbf{c}_\alpha u^{(0)}_{\alpha,1x1x}\,. \tag{3.3.29}$$

Using (3.3.25), one can find that $-\mathbf{C}_{\beta \mathrm{B}} + k_{\beta \mathrm{B}} = M^{*(-3)}_{\mathrm{B}\beta} - M^{*(-3)}_{\beta \mathrm{B}} = M^{*(-3)}_{\mathrm{B}\beta} s_{\mathrm{B}}$ and rewrite (3.3.28) and (3.3.29) in the form,

$$({}^1\!A_\beta V_{1,1x} + A^2_{\beta\alpha} u^{(0)}_{\alpha,1x1x} + B_\beta \phi_{,1x})_{,1x1x} \tag{3.3.30}$$

$$+\langle f_\beta\rangle + \langle g_\beta\rangle_S + (\langle f_1 y_\beta\rangle + \langle g_1 y_\beta\rangle_S)_{,1x}$$

$$= (M^{*(-3)}_{\mathrm{B}\beta} s_\beta u^{(0)}_{\mathrm{B},1x1x1x} + M^{*(-3)}_{1\beta} \phi_{,1x1x}) s_{\mathrm{B}} \text{ (no sum in } \beta, \mathrm{B} \text{ here)},$$

$$(A^1 V_{1,1x} + A_\alpha^2 u_{\alpha,1x1x}^{(0)} + B\phi_{,1x})_{,1x} \tag{3.3.31}$$

$$= (M_{31}^{*(-3)} u_{2,1x1x1x}^{(0)} + M_{21}^{*(-3)} u_{3,1x1x1x}^{(0)}) .$$

In the equations (3.3.30) and (3.3.31), $M_{\mathrm{B1}}^{*(-3)}$ means bending moments, and $M^{*(-3)} = M_{32}^{*(-3)} - M_{23}^{*(-3)}$ means the torsion moment corresponding to the initial stresses.

3.4 1-D Boundary Conditions Derived from 3-D Elasticity Problem

The 1-D model above was derived for a beam fastened along its ends $S_u = \{\mathbf{x} \in \partial G_\varepsilon : x_1 = \pm a\}$. We consider another type of boundary conditions interesting from a practical point of view and derive the corresponding boundary conditions. Let a beam be fastened along the surface $S_u = \{\mathbf{x} \in \partial G_\varepsilon : x_1 = -a\}$ (the left end of the beam), and at the surface $S_f = \{\mathbf{x} \in \partial G_\varepsilon : x_1 = a\}$ (the right end of the beam) let normal stresses $\mathbf{G}^\gamma(\mathbf{x}) = \varepsilon^{-2}\mathbf{g}^\gamma(\mathbf{x}) + \varepsilon^{-3}\mathbf{h}^\gamma(\mathbf{x})$ be applied to the beam (see Fig. 3.3).

We denote

$$\langle \circ \rangle_{S_f} = \int_{S_f} \circ \, \mathrm{d}\mathbf{x} = \varepsilon^{-1} \int_{\varepsilon^{-1} S_f} \circ \, \mathrm{d}\mathbf{y}$$

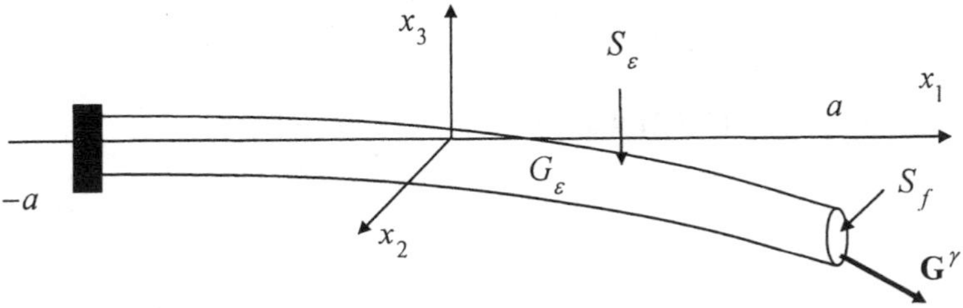

Fig. 3.3. Beam with a free edge

It will be assumed that

$$\langle \mathbf{g}^{\gamma}(\mathbf{x}) y_{\beta} \rangle_{S_f} = 0 \ (\beta = 2,3), \ \ \langle \mathbf{h}^{\gamma}(\mathbf{x}) \rangle_{S_f} = 0 .$$

This means that we decompose normal stresses $\mathbf{G}^{\gamma}$ into a sum, where $\varepsilon^{-2}\mathbf{g}^{\gamma}$ generates non-zero resultant forces and has zero moments and $\varepsilon^{-3}\mathbf{h}^{\gamma}$ has zero resultant forces and generates nonzero moments.

We denote $S_{\sigma} = S_{\varepsilon} \cup S_{f}$, where S_{ε} is the lateral surface of the plate, considered as a 3-D body; see Fig.3.3.

As above,

$$U_{\varepsilon} = \{\mathbf{v} \in \{H^1(G_{\varepsilon})\}^3 \ : \ \mathbf{v}(\mathbf{x}) = 0 \ \text{on} \ S_u \}$$

is the set of admissible displacements of the beam considered as a 3-D body.

For simplicity, we take the mass $\mathbf{f}^{\varepsilon}$ and surface $\mathbf{g}^{\varepsilon}$ forces equal to zero (it does not restrict the generality of our consideration). We consider a weak form of problem (3.2.11). In the case under consideration, it has the following form:

$$\int_{G_{\varepsilon}} \sigma_{ij} \frac{\partial v_i}{\partial x_j} \mathbf{dx} = \int_{S_f} \mathbf{G}^{\gamma} \mathbf{v} \mathbf{dx} \ \text{for any} \ \ \mathbf{v} \in U_{\varepsilon} . \tag{3.4.1}$$

Substituting expansion (3.2.8) in equilibrium equations (3.4.1), we obtain, with allowance for (3.2.9),

$$\sum_{p=-3}^{\infty} \int_{G_{\varepsilon}} \varepsilon^{p+2} \sigma_{ij}^{(p)} (v_{i,1x} \delta_{j1} + \varepsilon^{-1} v_{i,jy}) \mathbf{dx} \tag{3.4.2}$$

$$= \int_{S_f} \mathbf{G}^{\gamma} \mathbf{v} \mathbf{dx} \ \ \text{for any} \ \mathbf{v} \in U_{\varepsilon} .$$

We begin the sum in (3.4.2) with $p = -3$ because $\sigma_{ij}^{(-4)} = 0$; see (3.2.39) and (3.3.6).

As above,

$$H = \{\mathbf{w}(x_1) \in \{H^1([-a,a])\}^3 \ : \ \mathbf{w}(-a) = \mathbf{w}(a) = 0\} \subset U_{\varepsilon} .$$

We set $\mathbf{v} = \mathbf{w}(x_1) \in H$ in (3.4.2). Then, (3.4.2) takes the following form:

$$\sum_{p=-3}^{\infty} \int_{G_\varepsilon} \varepsilon^p \sigma_{i1}^{(p)} w_{i,1x} \mathbf{dx} = \int_{S_f} (\varepsilon^{-2}\mathbf{g}^\gamma + \varepsilon^{-3}\mathbf{h}^\gamma)\mathbf{w}\mathbf{dx} \quad \text{for any } \mathbf{w} \in H .$$

By virtue of Proposition 3.2 and condition $\langle \mathbf{h}^\gamma(\mathbf{x})\rangle_{S_f} = 0$, we find from this equation that

$$\sum_{p=-3}^{\infty} \int_{-a}^{a} \varepsilon^{p+2} N_{i1}^{(p)} w_{i,1x} \mathrm{d}x_1 = \mathbf{w}(a)\langle \mathbf{g}^\gamma \rangle_{S_f} \quad \text{for any } \mathbf{w} \in H . \tag{3.4.3}$$

Integrating (3.4.3) by parts, we obtain

$$\sum_{p=-3}^{\infty} [-\int_{-a}^{a} \varepsilon^{p+2} N_{i1,1x}^{(p)} w_i \mathrm{d}x_1 + \varepsilon^{p+2} N_{i1}^{(p)}(a) w_i(a)] \tag{3.4.4}$$

$$= \mathbf{w}(a)\langle \mathbf{g}^\gamma \rangle_{S_f} \quad \text{for any } \mathbf{w} \in H .$$

From (3.4.4), we obtain the equilibrium equations (3.2.15), (3.2.16) and the following boundary conditions:

$$N_{11}^{(-3)}(a) = 0 , \tag{3.4.5}$$

$$N_{i1}^{(-2)}(a) = \langle g_i^\gamma \rangle_{S_f} .$$

Now we set $\mathbf{v} = \mathbf{w}(x_1) y_\alpha$, where $\mathbf{w} \in H$ in (3.4.2) (it is clear that $\mathbf{w}(x_1) y_\alpha \in U_\varepsilon$). Then, (3.4.2) takes the following form:

$$\sum_{p=-3}^{\infty} \int_{G_\varepsilon} \varepsilon^p \sigma_{ij}^{(p)} (w_{i,1x}\delta_{j1} y_\alpha + \varepsilon^{-1} w_i \delta_{j\alpha}) \mathbf{dx}$$

$$= \int_{S_f} (\varepsilon^{-2}\mathbf{g}^\gamma + \varepsilon^{-3}\mathbf{h}^\gamma) y_\alpha \mathbf{w}\mathbf{dx} \quad \text{for any } \mathbf{w} \in H .$$

By virtue of Proposition 3.2 and condition $\langle \mathbf{g}^\gamma(\mathbf{x}) y_\beta \rangle_{S_f} = 0$ ($\beta = 2, 3$), we find from this equation that

$$\sum_{p=-3}^{\infty} \int_{-a}^{a} \varepsilon^{p+2}(\langle\sigma_{i1}^{(p)} y_\alpha\rangle w_{i,1x} + \varepsilon^{-1}\langle\sigma_{i\alpha}^{(p)}\rangle w_i)dx_1 = \varepsilon^{-1}\mathbf{w}(a)\langle\mathbf{h}^\gamma y_3\rangle_{S_f} \tag{3.4.6}$$

for any $\mathbf{w} \in H$.

Substituting $M_{i\beta}^{(p)}(x_1) = \langle\sigma_{j1}^{(p)}(x_1, \mathbf{y})y_\beta\rangle$ in (3.4.6) and integrating by parts, we obtain

$$\sum_{p=-3}^{\infty} [-\int_{-a}^{a} \varepsilon^{p+2}(M_{i\alpha,1x}^{(p)} + \varepsilon^{-1}\langle\sigma_{i\alpha}^{(p)}\rangle)w_i dx_1 + M_{i\alpha,1x}^{(p)} w_i(a)] \tag{3.4.7}$$

$$= \varepsilon^{-1}\mathbf{w}(a)\langle\mathbf{h}^\gamma y_3\rangle_{S_f} \quad \text{for any } \mathbf{w} \in H.$$

From (3.4.7), we obtain the equilibrium equations (3.2.18), (3.2.19) and the following boundary conditions:

$$M_{i\beta}^{(-3)}(a) - N_{i\beta}^{(-2)}(a) = \langle h_i^\gamma y_\beta\rangle_{S_f}. \tag{3.4.8}$$

We do not obtain here an expression for the edge normal forces through the normal deflections. As in the classical case, it can be expressed through the derivatives of moments and, thus, through normal deflections.

3.5 3-D and 1-D "Energy Forms" for a Stressed Beam and a Stability Criterion for a Beam

In this section the "energy form" (Trefftz, 1930, 1933; Marguerre, 1938; Prager, 1947; and Hill, 1958) for 3-D small diameter elastic body with initial stresses will be related to the "energy form" for a 1-D beam.

By establishing a relationship between the "energy form" for a 3-D small diameter elastic body (a real inhomogeneous 3-D beam) and an "energy form" for a 1-D beam, we will establish the relationship between the stability of 3-D and 1-D beams.

We define the "energy form" as

$$E(\sigma, \mathbf{u}) = \frac{1}{2} \int_{Q_\varepsilon} \sigma_{ij} \frac{\partial u_i^\varepsilon}{\partial x_j} \mathrm{d}\mathbf{x},$$

where

$$\sigma_{ij} = h_{ijmn}(x_1, x_2, \mathbf{x}/\varepsilon) \frac{\partial u_i^\varepsilon}{\partial x_j}$$

are local stresses. The coefficients h_{ijmn} are determined in (3.2.3).

Asymptotic of the "energy form" in a small-diameter domain

In accordance with differentiating rule (3.2.9), we can write

$$2E(\sigma, \mathbf{u}) = \int_{G_\varepsilon} \sigma_{ij}(u_{i,1x}^\varepsilon \delta_{j1} + \varepsilon^{-1} u_{i,jy}^\varepsilon) \mathrm{d}\mathbf{x}. \tag{3.5.1}$$

Substituting the asymptotic expansion for displacements (3.2.7) and the asymptotic expansion for stresses (3.2.8) in (3.5.1),

$$2E(\sigma, \mathbf{u}) = \int_{G_\varepsilon} \sum_{p=-3}^{\infty} \sum_{k=0}^{\infty} \varepsilon^{p+k} \sigma_{ij}^{(p)}(u_{i,1x}^{(k)} \delta_{j1} + \varepsilon^{-1} u_{i,jy}^{(k)}) \mathrm{d}\mathbf{x}. \tag{3.5.2}$$

Here, we take into account that $\sigma_{ij}^{(-4)} = 0$, in accordance with (3.2.39), and begin the expansion for stresses with $p = -3$. The leading terms in (3.5.2) are the following:

$$2E(\sigma,\mathbf{u}) = \int\limits_{G_\varepsilon} \varepsilon^{-3}\sigma_{ij}^{(-3)}(u_{i,1x}^{(0)}\delta_{j1} + \varepsilon^{-1}u_{i,jy}^{(1)})\mathbf{dx} \tag{3.5.3}$$

$$+ \int\limits_{G_\varepsilon} \varepsilon^{-2}\sigma_{ij}^{(-3)}(u_{i,1x}^{(1)}\delta_{j1} + \varepsilon^{-1}u_{i,jy}^{(2)})\mathbf{dx}$$

$$+ \int\limits_{G_\varepsilon} \varepsilon^{-2}\sigma_{ij}^{(-2)}(u_{i,1x}^{(0)}\delta_{j1} + \varepsilon^{-1}u_{i,jy}^{(1)})\mathbf{dx} + (...) \ ,$$

where (…) means the lower terms (the terms with the multipliers ε^k with $k \geq -1$).

We consider the term $u_{i,1x}^{(0)}\delta_{j1} + \varepsilon^{-1}u_{i,jy}^{(1)}$. Taking into account that $u_1^{(0)}(x_1) = 0$ ($\alpha = 1,2$) [see (3.2.36), (3.2.37)] and substituting $\mathbf{u}^{(1)}$ in accordance with (3.2.38), we obtain

$$u_{i,1x}^{(0)}\delta_{j\alpha} + \varepsilon^{-1}u_{i,jy}^{(1)} = \delta_{i\alpha}u_{\alpha,1x}^{(0)}\delta_{j1} - \delta_{i1}\delta_{j\alpha}u_{\alpha,1x}^{(0)} + \delta_{i\gamma}\delta_{j\Gamma}s_\gamma\phi \ . \tag{3.5.4}$$

Then,

$$\sigma_{ij}^{(-3)}(u_{i,1x}^{(0)}\delta_{j1} + \varepsilon^{-1}u_{i,jy}^{(1)}) = (\sigma_{\alpha 1}^{(-3)} - \sigma_{1\alpha}^{(-3)})u_{\alpha,1x}^{(0)} + \sigma_{\gamma\Gamma}^{(-3)}\phi = 0 \tag{3.5.5}$$

because $\sigma_{ij}^{(-3)}$ are symmetrical with respect to indexes i and j. Note that this symmetry holds for the case considered in Sect. 3.2, as well as for the case considered in Sect. 3.3. Allowing for (3.5.4), (3.5.3) takes the form,

$$2E(\sigma,\mathbf{u}) = \int\limits_{G_\varepsilon} \varepsilon^{-2}\sigma_{ij}^{(-3)}(u_{i,1x}^{(1)}\delta_{j1} + \varepsilon^{-1}u_{i,jy}^{(2)})\mathbf{dx} \tag{3.5.6}$$

$$+ \int\limits_{G_\varepsilon} \varepsilon^{-2}\sigma_{ij}^{(-2)}(u_{i,1x}^{(0)}\delta_{j1} + \varepsilon^{-1}u_{i,jy}^{(1)})\mathbf{dx} \ + (...).$$

We consider the last integral in (3.5.6). With regard to (3.5.5) and Proposition 3.2, it is equal to

$$\int\limits_{G_\varepsilon} \varepsilon^{-2}\sigma_{ij}^{(-2)}(u_{i,1x}^{(0)}\delta_{j1} + u_{i,jy}^{(1)})\mathbf{dx}$$

$$= \int_{G_\varepsilon} \varepsilon^{-2}[(\sigma_{\alpha 1}^{(-2)} - \sigma_{1\alpha}^{(-2)})u_{\alpha,1x}^{(0)} + \sigma_{\gamma\Gamma}^{(-2)} s_\gamma \phi]\mathbf{dx}$$

$$= \int_{-a}^{a} [(N_{\alpha 1}^{(-2)} - N_{1\alpha}^{(-2)})u_{\alpha,1x}^{(0)} + N_{\gamma\Gamma}^{(-2)} s_\gamma \phi]\mathrm{d}x_1 .$$

Note that $N_{32}^{(-2)} = N_{23}^{(-2)}$; see the comment on (3.2.66) for a beam under the action of initial axial force and equalities (3.3.20)–(3.3.23) for a beam under the action of moments of initial stresses. Then, the last integral is equal to

$$\int_{-a}^{a} (N_{\alpha 1}^{(-2)} - N_{1\alpha}^{(-2)})u_{\alpha,1x}^{(0)} \mathrm{d}x_1 .$$

In accordance with (3.2.61)–(3.2.66), (3.3.17), and (3.3.20), this integral is equal to

$$\int_{-a}^{a} (K_\alpha u_{\alpha,1x}^{(0)} - K_\alpha^m u_{\alpha,1x}^{(0)})\mathrm{d}x_1 \tag{3.5.7}$$

$$= \int_{-a}^{a} [N_{11}^{*}(u_{\alpha,1x}^{(0)})^2 + k_{\alpha\beta} u_{\alpha,1x1x}^{(0)} u_{\alpha,1x}^{(0)} + k_\beta u_{\alpha,1x}^{(0)} \phi_{,1x}]\mathrm{d}x_1 .$$

We consider the first integral in (3.5.6):

$$\int_{G_\varepsilon} \varepsilon^{-2} \sigma_{ij}^{(-3)} (u_{i,1x}^{(1)} \delta_{j1} + \varepsilon^{-1} u_{i,jy}^{(2)})\mathbf{dx} . \tag{3.5.8}$$

From (3.2.38), it follows that

$$u_{i,1x}^{(1)} = -y_\alpha \delta_{i1} u_{\alpha,1x1x}^{(0)}(x_1) + y_\Gamma s_\gamma \delta_{i\gamma} \phi_{,1x} + V_{i,1x}(x_1), \tag{3.5.9}$$

and from (3.2.43),

$$u_{i,jy}^{(2)} = X_{i,jy}^{1\alpha}(\mathbf{y}) u_{\alpha,1x1x}^{(0)}(x_1) - \delta_{j\alpha} V_{\alpha,1x}(x_1)\delta_{i1} + X_{i,jy}^{01}(\mathbf{y}) V_{1,1x}(x_1) \tag{3.5.10}$$

$$+ X_{i,jy}^{3}(\mathbf{y})\phi_{,1x}(x_1) + \delta_{j1} y_\Gamma s_\gamma \delta_{i\gamma} \psi(x_1) .$$

Taking into account (3.5.9) and (3.5.10), we can write

$$u^{(1)}_{i,1x}\delta_{j1}+u^{(2)}_{i,jy}=[-y_{\alpha}\delta_{i1}\delta_{j1}+X^{1\alpha}_{i,jy}(\mathbf{y})]u^{(0)}_{\alpha,1x1x}(x_1) \tag{3.5.11}$$

$$+[\delta_{i1}+X^{01}_{i,jy}(\mathbf{y})]V_{1,1x}(x_1)$$

$$+[\delta_{j1}y_{\Gamma}s_{\gamma}\delta_{i\gamma}+X^{3}_{i,jy}(\mathbf{y})]\phi_{,1x}(x_1)$$

$$+(\delta_{j1}\delta_{i\alpha}-\delta_{j\alpha}\delta_{i1})V_{\alpha,1x}+\delta_{j\Gamma}s_{\gamma}\delta_{i\gamma}\psi(x_1).$$

The expression

$$\sigma^{(-3)}_{ij}[(\delta_{j1}\delta_{i\alpha}-\delta_{j\alpha}\delta_{i1})V_{\alpha,1x}+\delta_{j\Gamma}s_{\gamma}\delta_{i\gamma}\psi(x_1)] \tag{3.5.12}$$

$$=(\sigma^{(-3)}_{\alpha 1}-\sigma^{(-3)}_{1\alpha})V_{\alpha,1x}+\sigma^{(-3)}_{\gamma\Gamma}s_{\gamma}\psi(x_1)$$

$$=(\sigma^{(-3)}_{\alpha 1}-\sigma^{(-3)}_{1\alpha})V_{\alpha,1x}+(\sigma^{(-3)}_{23}-\sigma^{(-3)}_{32})\psi(x_1)=0$$

due to the symmetry of $\sigma^{(-3)}_{ij}$.

Using (3.5.11), (3.5.12) and substituting $\sigma^{(-3)}_{ij}$ in accordance with (3.2.44), we write the integral (3.5.8) in the form

$$\int_{G_\varepsilon}\varepsilon^{-2}\sigma^{(-3)}_{ij}(u^{(1)}_{i,1x}\delta_{j1}+\varepsilon^{-1}u^{(2)}_{i,jy})d\mathbf{x} \tag{3.5.13}$$

$$=\int_{G_\varepsilon}\varepsilon^{-2}\{[c_{ijmn}(\mathbf{y})X^{01}_{m,ny}(\mathbf{y})-c_{ij11}(\mathbf{y}))V_{1,1x}(x_1)$$

$$+[c_{ijmn}(\mathbf{y})X^{1\alpha}_{m,ny}(\mathbf{y})+c_{ij11}(\mathbf{y})y_{\alpha}]u^{(0)}_{\alpha,1x1x}(x_1)$$

$$+[c_{ijmn}(\mathbf{y})X^{3}_{m,ny}(\mathbf{y})+c_{ij\gamma 1}(\mathbf{y})s_{\gamma}y_{\Gamma}]\phi_{,1x}(x_1)\}\times(u^{(1)}_{i,1x}\delta_{j1}+u^{(2)}_{i,jy})d\mathbf{x}$$

$$+\int_{G_\varepsilon}\varepsilon^{-2}\{[b_{ij\alpha 1}(\mathbf{y})-\mathrm{b}_{ij1\alpha}(\mathbf{y})]u^{(0)}_{\alpha,1x}(x_1)+b_{ij\gamma\Gamma}(\mathbf{y})s_{\gamma}\phi(x_1)\}$$

$$\times (u^{(1)}_{i,1x}\delta_{j1} + \varepsilon^{-1} u^{(2)}_{i,jy})\mathrm{d}\mathbf{x}\,.$$

We consider the last integral from (3.5.13). It can be written as

$$\int\limits_{G_\varepsilon} \varepsilon^{-2}\{[b_{ij\alpha 1}(\mathbf{y}) - \mathrm{b}_{ij1\alpha}(\mathbf{y})]u^{(0)}_{\alpha,1x}(x_1) + b_{ij\gamma\Gamma}(\mathbf{y})s_\gamma\phi(x_1)\} \tag{3.5.14}$$

$$\times\{[-y_\beta\delta_{i1}u^{(0)}_{\beta,1x1x}(x_1) + y_{\mathrm{B}}s_\beta]\delta_{i\beta}\phi_{,1x} + V_{i,1x}(x_1) + \varepsilon^{-1}u^{(2)}_{i,jy}\}\mathrm{d}\mathbf{x}$$

$$= \int\limits_{-a}^{a} \langle\{[b_{ij\alpha 1}(\mathbf{y}) - \mathrm{b}_{ij1\alpha}(\mathbf{y})]u^{(0)}_{\alpha,1x}(x_1) + b_{ij\gamma\Gamma}(\mathbf{y})s_\gamma\phi(x_1)\}$$

$$\times\{[-y_\beta\delta_{i1}u^{(0)}_{\beta,1x1x}(x_1) + y_{\mathrm{B}}s_\beta]\delta_{i\beta}\phi_{,1x} + V_{i,1x}(x_1))\delta_{j1} + \varepsilon^{-1}u^{(2)}_{i,jy}\}\rangle\mathrm{d}x_1$$

$$= \int\limits_{-a}^{a} \langle\{[b_{ij\alpha 1}(\mathbf{y}) - \mathrm{b}_{ij1\alpha}(\mathbf{y})]u^{(0)}_{\alpha,1x}(x_1) + b_{ij\gamma\Gamma}(\mathbf{y})s_\gamma\phi(x_1)\}$$

$$\times\delta_{j1}[-y_\beta\delta_{i1}u^{(0)}_{\beta,1x1x}(x_1) + y_{\mathrm{B}}s_\beta\delta_{i\beta}\phi_{,1x}]\rangle\mathrm{d}x_1\,.$$

Here we use Proposition 3.1 and the equalities (see Propositions 3.5 and 3.6)

$$\langle b_{ijmn}(\mathbf{y})u^{(2)}_{i,jy}\rangle = 0\,,\ \langle b_{ijmn}(\mathbf{y})\rangle = 0\,.$$

The integrand in the last integral in (3.5.15) can be written as

$$\langle -[b_{11\alpha 1}(\mathbf{y}) - \mathrm{b}_{111\alpha}(\mathbf{y})]y_\beta\rangle u^{(0)}_{\alpha,1x}(x_1)u^{(0)}_{\beta,1x1x}(x_1) \tag{3.5.15}$$

$$+\langle[b_{\beta 1\alpha 1}(\mathbf{y}) - b_{\beta 11\alpha}(\mathbf{y})]y_{\mathrm{B}}\rangle s_\beta u^{(0)}_{\alpha,1x}(x_1)\phi_{,1x}(x_1)$$

$$+\langle b_{\beta 1\gamma\Gamma}(\mathbf{y})y_{\mathrm{B}}\rangle s_\gamma s_\beta\phi(x_1)\phi_{,1x}(x_1) - \langle b_{11\gamma\Gamma}(\mathbf{y})y_{\mathrm{B}}\rangle s_\gamma\phi(x_1)u^{(0)}_{\beta,1x1x}(x_1)$$

$$= -\mathbf{C}_{\beta\alpha}u^{(0)}_{\alpha,1x}(x_1)u^{(0)}_{\beta,1x1x}(x_1) + \mathbf{c}_\alpha u^{(0)}_{\alpha,1x}(x_1)\phi_{,1x}(x_1)\,.$$

Here, we use the notations (3.3.13) and the following equalities:

$$\langle b_{11\gamma\Gamma}(\mathbf{y})y_\beta\rangle = \delta_{1\gamma}\langle \sigma^*_{1\Gamma}(\mathbf{y})y_\beta\rangle = 0$$

(because $\delta_{1\gamma} = 0$, remember that α takes values 2 and 3), and the equalities

$$\langle b_{\beta 1\gamma\Gamma}(\mathbf{y})y_\beta\rangle s_\gamma s_\beta = \langle b_{21\gamma\Gamma}(\mathbf{y})y_3\rangle s_\gamma - \langle b_{31\gamma\Gamma}(\mathbf{y})y_2\rangle s_\gamma$$

$$= \langle \sigma^*_{13}(\mathbf{y})y_3\rangle - \langle \sigma^*_{12}(\mathbf{y})y_2\rangle = 0 .$$

The last equality holds by virtue of Proposition 3.6.

Allowing for (3.5.14) and (3.5.15), we find that the right-hand side of (3.4.13) is equal to

$$\int_{-a}^{a} \langle [c_{ijmn}(\mathbf{y})X^{01}_{m,ny}(\mathbf{y}) - c_{ij11}(\mathbf{y})]V_{1,1x}(x_1) \tag{3.5.16}$$

$$+ [c_{ijmn}(\mathbf{y})X^{1\alpha}_{m,ny}(\mathbf{y}) + c_{ij11}(\mathbf{y})y_\alpha]u^{(0)}_{\alpha,1x1x}(x_1)$$

$$+ [c_{ijmn}(\mathbf{y})X^{3}_{m,ny}(\mathbf{y}) + c_{ij\gamma 1}(\mathbf{y})s_\gamma y_\Gamma)\phi_{,1x}(x_1)]$$

$$\times [(-y_\alpha \delta_{i1}\delta_{j1} + X^{1\alpha}_{i,jy}(\mathbf{y})]u^{(0)}_{\alpha,1x1x}(x_1)$$

$$+ [\delta_{i1} + X^{01}_{i,jy}(\mathbf{y})]V_{1,1x}(x_1)$$

$$+ [y_\Gamma s_\gamma \delta_{i\gamma}\phi_{,1x} + X^{3}_{i,jy}(\mathbf{y})]\phi_{,1x}(x_1)\rangle \mathrm{d}x_1 .$$

As a result, we find that the leading term of the "energy form" $E(\sigma,\mathbf{u})$ for an inhomogeneous beam considered as a 3-D small-diameter body is of the order $\varepsilon^0 = 1$ and it is equal to the sum of (3.5.7), (3.5.15), and (3.5.16):

$$E(\sigma,\mathbf{u}) = E(u^{(0)}_2, u^{(0)}_3, V_1, \phi) \tag{3.5.17}$$

$$-\frac{1}{2}\int_{-a}^{a} \mathbf{C}_{\beta\alpha} u^{(0)}_{\alpha,1x}(x_1) u^{(0)}_{\beta,1x1x}(x_1)\mathrm{d}x_1$$

$$+\frac{1}{2}\int_{-a}^{a} \mathbf{c}_{\alpha} u_{\alpha,1x}^{(0)}(x_1)\phi_{,1x}(x_1)\mathrm{d}x_1$$

$$+\frac{1}{2}\int_{-a}^{a} [N_{11}^{*}(u_{\alpha,1x}^{(0)})^2 + k_{\alpha\beta} u_{\alpha,1x1x}^{(0)} u_{\alpha,1x}^{(0)} + k_{\beta} u_{\alpha,1x}^{(0)}\phi_{,1x}]\mathrm{d}x_1 \,.$$

Here, $E(u_2^{(0)}, u_3^{(0)}, V_1, \phi)$, the energy of a homogenized 1-D beam with no initial stresses, is computed in accordance with (3.5.17), which can be written as

$$E(u_2^{(0)}, u_3^{(0)}, V_1, \phi) \tag{3.5.18}$$

$$=\frac{1}{2}\int_{-a}^{a} \{A[V_{1,1x}(x_1)]^2 + A_{\alpha\beta}^2 u_{\alpha,1x1x}^{(0)} u_{\beta,1x1x}^{(0)} + B[\phi_{,1x}(x_1)]^2$$

$$+(A_{\alpha}^1 + {}^1A_{\alpha}) u_{\alpha,1x1x}^{(0)}(x_1) V_{1,1x}(x_1) + (b+B) u_{\alpha,1x1x}^{(0)}(x_1) V_{1,1x}(x_1)$$

$$+(B_{\alpha} + A_{\alpha z}^2) u_{\alpha,1x1z}^{(0)}(x_1)\phi_{1,1x}(x_1)\}\mathrm{d}x_1 \,.$$

The coefficients in (3.5.18) are determined by (3.2.48)

Representation (3.5.17) makes it possible to apply the homogenization method developed above for the analysis of the stability of plates of complex structure.

In the works by Hill (1958) and Prager (1947), a criterion of elastic stability was formulated in terms of "energy form." The criterion mentioned states the following: if the "energy form,"

$$E(\sigma, \mathbf{u}) = E(u_2^{(0)}, u_3^{(0)}, V_1, \phi)$$

$$-\frac{1}{2}\int_{-a}^{a} \mathbf{C}_{\beta\alpha} u_{\alpha,1x}^{(0)}(x_1) u_{\beta,1x1x}^{(0)}(x_1)\mathrm{d}x_1$$

$$+\frac{1}{2}\int_{-a}^{a} \mathbf{c}_{\alpha} u_{\alpha,1x}^{(0)}(x_1)\phi_{,1x}(x_1)\mathrm{d}x_1$$

$$+\frac{1}{2}\int_{-a}^{a}[N_{11}^{*}(u_{\alpha,1x}^{(0)})^{2}+K_{\alpha\beta}u_{\alpha,1x1x}^{(0)}u_{\alpha,1x}^{(0)}+k_{\beta}u_{\alpha,1x}^{(0)}\phi_{,1x}]\mathrm{d}x_{1}\geq 0$$

for any admissible displacements $u_{2}^{(0)}, u_{3}^{(0)}, V_{1}, \phi$, then the body is stable; if the "energy form,"

$$E(\sigma,\mathbf{u})=E(u_{2}^{(0)},u_{3}^{(0)},V_{1},\phi)$$

$$-\frac{1}{2}\int_{-a}^{a}\mathbf{c}_{\beta\alpha}u_{\alpha,1x}^{(0)}(x_{1})u_{\beta,1x1x}^{(0)}(x_{1})\mathrm{d}x_{1}$$

$$+\frac{1}{2}\int_{-a}^{a}\mathbf{c}_{\alpha}u_{\alpha,1x}^{(0)}(x_{1})\phi_{,1x}(x_{1})\mathrm{d}x_{1}$$

$$+\frac{1}{2}\int_{-a}^{a}[N_{11}^{*}(u_{\alpha,1x}^{(0)})^{2}+K_{\alpha\beta}u_{\alpha,1x1x}^{(0)}u_{\alpha,1x}^{(0)}+k_{\beta}u_{\alpha,1x}^{(0)}\phi_{,1x}]\mathrm{d}x_{1}$$

takes both positive and negative values, then the body is not stable, and there exist nonzero displacements $u_{2}^{(0)}, u_{3}^{(0)}, V_{1}, \phi$, which satisfy the equations

$$\delta_{V_{1}}E(u_{2}^{(0)},u_{3}^{(0)},V_{1},\phi)=0, \qquad (3.5.19)$$

$$\delta_{u_{\alpha}}E(u_{2}^{(0)},u_{3}^{(0)},V_{1},\phi)=0,\ (\alpha=1,2),$$

$$\delta_{\phi}E(u_{2}^{(0)},u_{3}^{(0)},V_{1},\phi)=0,$$

where δ means the variation of functional $E(u_{2}^{(0)},u_{3}^{(0)},V_{1},\phi)$ (3.5.17) with respect to the corresponding function.

The buckling problem

Let us write (3.5.19) for the special case, when the additional forces are equal to zero: $\mathbf{f}^{\varepsilon}=0$, g=0; and the initial stresses are proportional to a parameter λ and satisfy the conditions of Proposition 3.5. The variational equalities (3.5.19) can be written in the form of the following equations:

$$(AV_{1,1x} + A^1_\alpha u^{(0)}_{\alpha,1x} + b\phi_{,1x})_{,1x} = 0\,, \tag{3.5.20}$$

$$({}^1A_\beta V_{1,1x} + A^2_{\beta\alpha} u^{(0)}_{\alpha,1x1x} + B_\beta \phi_{,1x})_{,1x1x} \tag{3.5.21}$$

$$= \lambda[(N^*_{11} u^{(0)}_{\alpha,1x})_{,1x} - (\boldsymbol{C}_{\beta\mathrm{B}} u^{(0)}_{\mathrm{B},1x})_{,1x1x} + (k_{\beta\mathrm{B}} u^{(0)}_{\mathrm{B},1x1x} + k_\beta \phi_{,1x})_{,1x}]\,,$$

$$(A^1 V_{1,1x} + A^2_\alpha u^{(0)}_{\alpha,1x1x} + B\phi_{,1x})_{,1x} = \lambda(\boldsymbol{c}_\alpha u^{(0)}_{\alpha,1x})_{,1x}\,, \tag{3.5.22}$$

with the boundary conditions

$$V_1(-a) = V_1(a) = 0, \tag{3.5.23}$$

$$u^{(0)}_\alpha(-a) = u^{(0)}_\alpha(a) = u^{(0)}_{\alpha,1x}(-a) = u^{(0)}_{\alpha,1x}(a) = 0,$$

$$\phi(-a) = \phi(a) = 0\,.$$

Note that

$$\boldsymbol{C}_{\beta\mathrm{B}} = M^{*(-3)}_{\beta\mathrm{B}}\,,$$

$$\boldsymbol{c}_\alpha = M^{*(-3)}_{12}\delta_{3\alpha} - M^{*(-3)}_{13}\delta_{2\alpha}\,,$$

$$k_{\beta\mathrm{B}} = -M^{*(-3)}_{\beta\mathrm{B}}\,,$$

$$k_\beta = M^{*(-3)}_{1\beta} s_\mathrm{B} \quad \text{(no sum in } \beta, \mathrm{B} \text{ here)}$$

are the moments of the initial stresses.

3.6 Strings (1-D Model)

The problem involving inhomogeneous strings (cables, wire ropes, and so on), in which both the inhomogeneity of structure and the initial stresses play a decisive role, is directly related to the problems considered in the book.

We will term a string a body of small diameter whose bending stiffnesses can be neglected when we study those normal deflections. The deflections of a string are determined by the initial axial force.

For strings, whose diameter is significantly less then the length of the period, a 1-D model of a string may be taken as the initial equation. In this section, we present the homogenization procedure for a 1-D model of an inhomogeneous string.

We will examine a 1-D model of a string of periodic structure obtained by repeating a certain small periodicity cell over the Ox_1 axis (Fig. 3.4). Here, ε is the periodicity cell's characteristic dimension, which is assumed to be small (that is formalized in the form $\varepsilon \to 0$).

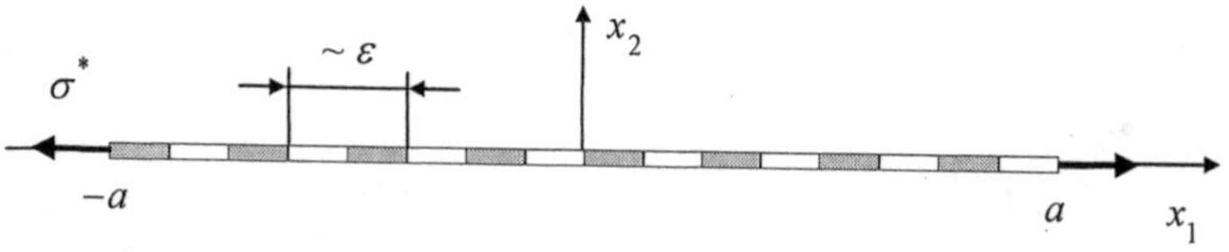

Fig. 3.4. A 1-D string of periodic structure

The starting point of our investigation is the 1-D formulation of a string problem. In accordance with this model, the equilibrium equations of the string can be written in the form,

$$\frac{dM}{dx} = f\,(x, x/\varepsilon) \quad \text{in } G, \tag{3.6.1}$$

where

$$Mw^{\varepsilon} = \sigma^* \frac{dw^{\varepsilon}}{dx}. \tag{3.6.2}$$

The boundary conditions can be written in the form,

$$w^{\varepsilon}(-a) = w^{\varepsilon}(a) = 0. \tag{3.6.3}$$

Here, $-a$ and a are the coordinates of the ends of the string (see Fig. 3.4), w^{ε} is the normal deflection, σ^* is the initial axial force, and f is the mass force.

In the case under consideration, the axial stress σ^* is a constant; see Sect. 3.1. The limit problem is obtained in a simple way, and it has the form,

$$\sigma^* w^{(0)}_{,xx} = \langle f \rangle(x), \tag{3.6.4}$$

$$w^{(0)}(-a) = w^{(0)}(a) = 0 \,. \qquad (3.6.5)$$

3.7 Strings (Transition from a 3-D to a 1-D Model)

In this section, a 1-D string model will be derived directly from the 3-D elasticity problem. While there is a certain resemblance between the methods used before and those used in this section, the results obtained do not follow directly from the results presented before.

Statement of the problem

We consider a domain G_ε that is obtained by periodic repetition of a basic periodicity cell εY on the Ox_1 axis, see Fig. 3.5. The characteristic size of this periodicity cell (which is similar to the characteristic diameter of a string) is a small quantity $\varepsilon << 1$; see Fig. 3.5. The condition $\varepsilon << 1$ is formalized in the form $\varepsilon \to 0$. When $\varepsilon \to 0$, domain G_ε "tightens" to the interval $[-a, a]$ on the Ox_1 axis. We assume that the structure under consideration has initial stresses σ^*_{ij} and these stresses are determined from the solution of a problem of the theory of elasticity.

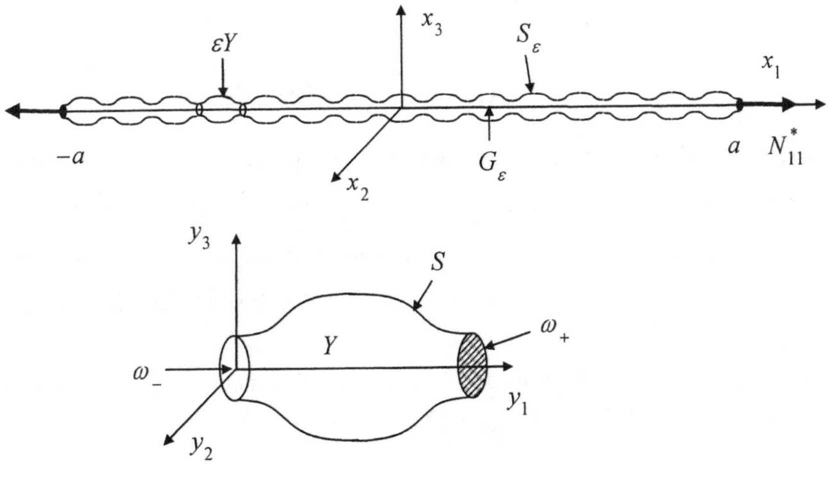

Fig. 3.5. An inhomogeneous string and its periodicity cell

In classical theory, stiffness of string is neglected compared with initial stresses. We shall assume that the elasticity constants of the string material c_{ijkl} and the initial stresses σ^*_{ij} are of the same order. The reason for this choice is that it is impossible to create initial stresses σ^*_{ij} of a greater order than the order of the elastic constants c_{ijkl}. Thus, in a rigorous theory, we cannot neglect the stiffnesses of the string. The rigorous theory must not neglect the stiffnesses but explains why the stiffnesses do not influence the behavior of strings.

The starting point for our investigation is a 3-D model of an elastic body with initial stresses (see Sect. 1.1). The equilibrium equations have the form,

$$\frac{\partial \sigma_{ij}}{\partial x_j} + \varepsilon^{-2} f(x_1, \mathbf{x}/\varepsilon) = 0 \quad \text{in } G_\varepsilon, \tag{3.7.1}$$

$$\sigma_{ij} n_j = \varepsilon^{-1} g(x_1, \mathbf{x}/\varepsilon) \quad \text{on } S_\sigma ,$$

$$\mathbf{u}^\varepsilon = 0 \quad \text{on } S_u = \{ \mathbf{x} \in G_\varepsilon : x_1 = \pm a \},$$

where S_σ and S_u are surfaces of domain G_ε (see Fig. 3.5).

We write the relation between current stresses σ_{ij}, displacements $\mathbf{u}^\varepsilon$, and initial stresses σ^*_{ij} in the following form:

$$\sigma_{ij} = \varepsilon^{-2} h_{ijmn}(x_1, \mathbf{x}/\varepsilon) \frac{\partial u^\varepsilon_m}{\partial x_n}, \tag{3.7.2}$$

$$h_{ijmn}(x_1, \mathbf{x}/\varepsilon) = c_{ijmn}(\mathbf{x}/\varepsilon) + \sigma^*_{jn}(x_1, \mathbf{x}/\varepsilon)\delta_{im}, \tag{3.7.3}$$

where c_{ijmn} are the components of the elastic moduli tensor, σ^*_{jn} are the components of the tensor of initial stresses, and x_1 is the "slow" variable along the axis of the string. Functions $c_{ijmn}(\mathbf{y})$, $\sigma^*_{jn}(x_1, \mathbf{y})$ are periodic with respect to y_1 with period T (T is the projection of the cell Y on the Oy_1 axis in "fast" variable $\mathbf{y} = \mathbf{x}/\varepsilon$; Fig. 3.5).

Equation (3.7.2) can be considered the governing equation of a stressed body. The symmetry of coefficients h_{ijmn} is described in Proposition 1.1.

The multiplier ε^{-2} in (3.7.2) guarantees that the axial stiffness of the string will be nonzero, as $\varepsilon \to 0$. The term $\varepsilon^{-2}\sigma^{*}_{jn}$ corresponds to the axial load of the string, which is independent of ε. The multiplier ε^{-2} in the term $\varepsilon^{-2}\mathbf{f}$ guarantees that the mass force is nonzero, as $\varepsilon \to 0$. The multiplier ε^{-1} in the term $\varepsilon^{-1}\mathbf{g}$ guaranties that the surface force be nonzero, as $\varepsilon \to 0$. Therefore, the powers of ε in (3.7.1) and (3.7.2) correspond to the order of the forces and stiffness estimated from the physical point of view.

Asymptotic expansions for string

We consider "slow" variable x_1 along the axis of the string and "fast" variable $\mathbf{y} = \mathbf{x}/\varepsilon$. The derivative of a function of the form, $Z(x_1, \mathbf{y})$, is calculated by replacing the differential operator according to the rule,

$$\frac{\partial Z}{\partial x_1} = Z_{,1x} + \varepsilon^{-1} Z_{,1y}, \tag{3.7.4}$$

$$\frac{\partial Z}{\partial x_\alpha} = \varepsilon^{-1} Z_{,\alpha y} \quad (\alpha = 2, 3).$$

The Latin indexes take values 1,2,3, and the Greek indexes take values 2,3; subscript $,1x$ means $\partial/\partial x_1$, and subscript $,iy$ means $\partial/\partial y_i$.

We shall seek a solution of problem (3.7.1)–(3.7.3) in a form, which is analogous to that used previously, but in accordance with (3.7.2), we start the expansion for stresses σ_{ij} with a term of the order of ε^{-2}:

$$\mathbf{u}^{\varepsilon} = \mathbf{u}^{(0)}(x_1) + \varepsilon \mathbf{u}^{(1)}(x_1, \mathbf{y}) + \ldots = \mathbf{u}^{(0)}(x_1) + \sum_{k=1}^{\infty} \varepsilon^{k} \mathbf{u}^{(k)}(x_1, \mathbf{y}), \tag{3.7.5}$$

$$\sigma_{ij} = \sum_{p=-2}^{\infty} \varepsilon^{p} \sigma_{ij}^{(p)}(x_1, \mathbf{y}). \tag{3.7.6}$$

All the functions on the right–hand sides of (3.7.5), (3.7.6) are assumed to be periodic in x_1 with period T.

Substituting (3.7.5) in (3.7.2) and taking into account differentiating rule (3.7.4),

$$\sum_{p=-2}^{\infty} \varepsilon^{p} \sigma_{ij}^{(p)} = \sum_{k=0}^{\infty} \varepsilon^{k-2} h_{ijm1}(x_1, \mathbf{y}) u^{(k)}_{m,1x} + \varepsilon^{k-3} h_{ijmn}(x_1, \mathbf{y}) u^{(k)}_{m,ny}. \tag{3.7.7}$$

Equating the expressions in (3.7.7) with the identical powers of ε, we obtain

$$\sigma_{ij}^{(p)} = h_{ijmn}(x_1, \mathbf{y}) u_{m,ny}^{(p+3)} + h_{ijm1}(x_1, \mathbf{y}) u_{m,1x}^{(p+2)} \quad (p = -2, -1, \ldots). \qquad (3.7.8)$$

We consider equilibrium equation (3.7.1). By substituting (3.7.6) in (3.7.1) and taking into account (3.7.4), we find

$$\sum_{p=-2}^{\infty} \varepsilon^p (\varepsilon^{-1} \sigma_{ij,jy}^{(p)} + \sigma_{i1,1x}^{(p)}) + \varepsilon^{-2} f_i(x_1, \mathbf{y}) = 0 \quad \text{in } Y,$$

$$\sum_{p=-2}^{\infty} \varepsilon^p \sigma_{ij}^{(p)} n_j = \varepsilon^{-1} g_i(x_1, \mathbf{y}) \quad \text{on } S.$$

Here, S is the lateral surface of periodicity cell Y (Fig. 3.5).

Equating the terms accompanying the same powers of ε, we obtain the following equations:

$$\sigma_{ij,jy}^{(p)} + \sigma_{i1,1x}^{(p)} = \begin{cases} 0 \text{ if } p \neq -1, \\ -f_i \text{ if } p = -1, \end{cases} \quad \text{in } Y; \qquad (3.7.9)$$

$$\sigma_{ij}^{(p)} n_j = \begin{cases} 0 \text{ if } p \neq -1, \\ g_i \text{ if } p = -1, \end{cases} \quad \text{on } S.$$

We determine, that $\sigma_{ij}^{(-3)} = 0$.

Only the case $p = -2$ will be interesting for us. For this case, equality (3.7.8) takes the form,

$$\sigma_{ij}^{(-2)} = h_{ijmn}(x_1, \mathbf{y}) u_{m,ny}^{(1)} + h_{ijm1}(x_1, \mathbf{y}) u_{m,1x}^{(0)}(x_1), \qquad (3.7.10)$$

and (3.7.9) takes the form

$$\begin{cases} \sigma_{ij,jy}^{(-2)} = 0 \quad \text{in } Y, \\ \sigma_{ij}^{(-2)} n_j = 0 \quad \text{on } S. \end{cases} \qquad (3.7.11)$$

Substituting (3.7.10) in (3.7.11), we obtain

$$\begin{cases} [h_{ijmn}(x_1,\mathbf{y})u^{(1)}_{m,ny} + h_{ijm1}(\mathbf{y})u^{(0)}_{m,1x}(x_1)]_{,jy} = 0 & \text{in } Y, \\ [h_{ijmn}(x_1,\mathbf{y})u^{(1)}_{m,ny} + h_{ijm1}(\mathbf{y})u^{(0)}_{m,1x}(x_1)]n_j = 0 & \text{on } S, \\ \mathbf{u}^{(0)}(\mathbf{y}) \text{ is periodic in } y_1 \text{ with period } T. \end{cases} \tag{3.7.12}$$

Then, we arrive in the usual way at the cellular problem for string

$$\begin{cases} [h_{ijmn}(x_1,\mathbf{y})K^p_{m,ny} + h_{ijp1}(x_1,\mathbf{y})]_{,jy} = 0 & \text{in } Y, \\ [h_{ijmn}(x_1,\mathbf{y})K^p_{m,ny} + h_{ijp1}(x_1,\mathbf{y})]n_j = 0 & \text{on } S, \\ \mathbf{K}^p(\mathbf{y}) \text{ is periodic in } y_1 \text{ with period } T. \end{cases} \tag{3.7.13}$$

As was suggested above, initial stresses σ^*_{ij} are determined from the solution of a problem in the theory of elasticity. Then,

$$\sigma^*_{ij,jy} = 0 \text{ in } Y, \tag{3.7.14}$$

$$\sigma^*_{ij}n_j = 0 \text{ on } S.$$

Cellular problem (3.7.13) looks similar to the cellular problem for a beam with no initial stresses (3.2.29). Nevertheless, these problems are significantly different. The difference is related to the coefficients of the problems. The coefficients h_{ijmn} in problem (3.7.13) depends on the initial stresses and do not have the symmetry occurring in elastic constants c_{ijmn} [which are the coefficients of the problem (3.2.29)]. The symmetry of coefficients c_{ijmn} plays the determining role in the proof of Proposition 3.4 (which introduces the generalizations of Bernoulli-Navier and Saint-Venant solutions). We demonstrate that an analog of Proposition 3.4 holds for a stressed beam under conditions (3.7.14) with respect to the initial stresses.

Proposition 3.8. *Under conditions* (3.7.14), *cellular problem* (3.7.13) *is degenerate. Its kernel is generated by the following two functions:* **const** $= (C_1, C_2, C_3)$ *and*

$$\mathbf{U}(\mathbf{y}) = y_\gamma s_\gamma \mathbf{e}_\Gamma, \tag{3.7.15}$$

($\Gamma=2$ if $\gamma=3$ and $\Gamma=3$ if $\gamma=2$; $s_2=1$, and $s_3=-1$).

Using the solution of cellular problem (3.7.13) and equality (3.7.15), we can write the following representation for the solution of problem (3.7.12):

$$\mathbf{u}^{(1)} = \mathbf{K}^{p}(\mathbf{y})u^{(0)}_{p,1x}(x_1) + \mathbf{V}(x_1) + y_\gamma s_\gamma \mathbf{e}_\Gamma \phi(x_1). \tag{3.7.16}$$

Here, we use the fact that functions of variable x_1 play the role of constants in problems in variables $\mathbf{y}$.

In accordance with Proposition 3.8, under conditions (3.7.14), the following equation holds:

$$\mathbf{K}^{\alpha}(\mathbf{y}) = -y_\alpha \mathbf{e}_1. \tag{3.7.17}$$

By (3.7.17), representation (3.7.16) takes the form,

$$\mathbf{u}^{(1)} = \mathbf{K}^{1}(\mathbf{y})u^{(0)}_{1,1x}(x_1) - y_\alpha \mathbf{e}_1 u^{(0)}_{\alpha,1x}(x_1) + \mathbf{V}(x_1) + y_\gamma s_\gamma \mathbf{e}_\Gamma \phi(x_1). \tag{3.7.18}$$

This representation is identical in form to representation (3.2.32), but $\mathbf{K}^1(\mathbf{y})$ in (3.7.18) depends on initial stresses.

Substituting of (3.7.18) in (3.7.10) gives

$$\sigma_{ij}^{(-2)} = [-h_{ij1\alpha}(x_1,\mathbf{y}) + h_{ij\alpha 1}(x_1,\mathbf{y})]u^{(0)}_{\alpha,1x}(x_1) \tag{3.7.19}$$

$$+[h_{ij11}(x_1,\mathbf{y}) + h_{ijmn}(x_1,\mathbf{y})K^{1}_{m,ny}(\mathbf{y})]u^{(0)}_{1,1x}(x_1)$$

$$+ h_{ij\Gamma\gamma}(x_1,\mathbf{y})s_\gamma \phi(x_1).$$

By using the definition of h_{ijkl} (3.7.2) and the symmetry of the elastic moduli (see condition *C1*, Sect. 1.2), we obtain from (3.7.19),

$$\sigma_{ij}^{(-2)} = [-\sigma^{*}_{j\alpha}(x_1, \mathbf{y})\delta_{i1} + \sigma^{*}_{j1}(x_1, \mathbf{y})\delta_{i\alpha}]u^{(0)}_{\alpha,1x}(x_1) \tag{3.7.20}$$

$$+[h_{ij11}(x_1, \mathbf{y}) + h_{ijmn}(x_1, \mathbf{y})K^{1}_{m,ny}(\mathbf{y})]u^{(0)}_{1,1x}(x_1)$$

$$+\sigma^{*}_{j\gamma}(x_1, \mathbf{y})\delta_{i\Gamma}s_{\gamma}\phi(x_1)\,.$$

1-D model of a string

We denote

$$\langle\circ\rangle = T^{-1}\int_{Y}\circ\,\mathbf{dy} \text{ and } \langle\circ\rangle_S = T^{-1}\int_{S}\circ\,\mathbf{dy}$$

as an average over basic cell Y and an average over lateral surface S of periodicity cell Y (Fig. 3.5).

1-D equations of equilibrium

Let us introduce the resultant forces $N_i = \langle\sigma^{(-2)}_{i1}\rangle$. In contrast to the classical theory, we introduce the forces by averaging the local stresses over the periodicity cell. In the classical theory, the forces are introduced by averaging the local stresses over the cross section of a string.

We derive the equilibrium equations for these internal forces. Equilibrium equations (3.7.1) can be written in the following form:

$$\int_{G_\varepsilon}\sigma_{ij}\frac{\partial v_i}{\partial x_j}\mathbf{dx} = \varepsilon^{-1}\int_{S_\varepsilon}\mathbf{gvdx} + \varepsilon^{-2}\int_{G_\varepsilon}\mathbf{fvdx} \quad \text{for any } \mathbf{v}\in U_\varepsilon, \tag{3.7.21}$$

where $\mathbf{v}$ is the trial function.

Substituting (3.7.6) in (3.7.21) and taking into account differentiating rule (3.7.4), we obtain

$$\sum_{p=-2}^{\infty}\int_{G_\varepsilon}\varepsilon^{p}(\varepsilon^{-1}\sigma^{(p)}_{ij}v_{i,jy} + \sigma^{(p)}_{i1}v_{i,1x})\mathbf{dx} \tag{3.7.22}$$

$$= \varepsilon^{-1}\int_{S_\varepsilon}\mathbf{gvdx} + \varepsilon^{-2}\int_{G_\varepsilon}\mathbf{fvdx} \quad \text{for any } \mathbf{v}\in U_\varepsilon\,.$$

We take the trial function in the form, $\mathbf{v} = \mathbf{w}(x_1) \in H$. Taking into account Proposition 3.2 [see formula (3.7.23)], we obtain the following equality from (3.7.22):

$$\sum_{p=-2}^{\infty} \int_{-a}^{a} \varepsilon^{p+2} \langle \sigma_{i1}^{(p)} \rangle w_{i,1x} dx_1 = \int_{-a}^{a} \langle \mathbf{g} \rangle_S \mathbf{w} dx_1 \tag{3.7.23}$$

$$+ \int_{-a}^{a} \langle f_\alpha \rangle w_\alpha dx_1 \quad \text{for any } \mathbf{w} \in H .$$

Integrating (3.7.23) by parts and equating the expressions corresponding to ε^0 [it is the leading term in (3.7.23)], we obtain the following 1-D equilibrium equations:

$$N_{1,1x} + \langle f_1 \rangle + \langle g_1 \rangle_S = 0 , \tag{3.7.24}$$

$$N_{\alpha,1x} + \langle f_\alpha \rangle + \langle g_\alpha \rangle_S = 0 . \tag{3.7.25}$$

1-D governing equations

Using equalities (3.7.20), we can express forces N_i through deformation characteristics $u_{i,1x}^{(0)}$ (i=1, 2, 3). Averaging (3.7.20) over periodicity cell Y,

$$N_i = \langle -\overset{*}{\sigma}_{1\alpha}(x_1, \mathbf{y})\delta_{i1} + \overset{*}{\sigma}_{11}(x_1, \mathbf{y})\delta_{i\alpha} \rangle u_{\alpha,1x}^{(0)}(x_1) \tag{3.7.26}$$

$$+ \langle h_{i111}(x_1, \mathbf{y}) + h_{i1mn}(x_1, \mathbf{y}) K_{m,ny}^1(\mathbf{y}) \rangle u_{1,1x}^{(0)}(x_1)$$

$$+ \langle \overset{*}{\sigma}_{1\gamma}(x_1, \mathbf{y}) \rangle \delta_{i\Gamma} s_\gamma \phi(x_1) .$$

Let us consider (3.7.26) for different values of indexes. For $i = 1$ (3.7.26), by using (3.7.3),

$$N_1 = -\langle \overset{*}{\sigma}_{1\alpha}(x_1, \mathbf{y}) \rangle u_{\alpha,1x}^{(0)}(x_1) \tag{3.7.27}$$

$$+ \langle h_{1111}(x_1, \mathbf{y}) + h_{11mn}(x_1, \mathbf{y}) K_{m,ny}^1(\mathbf{y}) \rangle u_{1,1x}^{(0)}(x_1)$$

since $\delta_{1\Gamma} = 0$.

For $i = \beta$, (3.7.26) and (3.7.3) give

$$N_\beta = -\langle \sigma^*_{11}(x_1, \mathbf{y}) \rangle u^{(0)}_{\beta,1x}(x_1) \tag{3.7.28}$$

$$+ \langle h_{\beta 111}(x_1, \mathbf{y}) + h_{\beta 1mn}(x_1, \mathbf{y}) K^1_{m,ny}(\mathbf{y}) \rangle u^{(0)}_{1,1x}(x_1)$$

$$+ \langle \sigma^*_{1\mathrm{B}}(x_1, \mathbf{y}) \rangle s_{\mathrm{B}} \phi(x_1) .$$

Let σ^*_{ij} be determined from the solution of elasticity theory problem (3.7.14). Then, in accordance with Proposition 3.6, $\langle \sigma^*_{i\alpha}(x_1, \mathbf{y}) \rangle = 0$. Then, the terms containing $\sigma^*_{1\mathrm{B}}$ in (3.7.28) vanish, and (3.7.27), (3.7.28) take the form,

$$N_1 = A(\sigma) u^{(0)}_{1,1x}(x_1) , \tag{3.7.29}$$

$$N_\beta = N^* u^{(0)}_{\beta,1x}(x_1) + R_\beta(\sigma) u^{(0)}_{1,1x}(x_1) , \tag{3.7.30}$$

where

$$A(\sigma) = \langle h_{1111}(x_1, \mathbf{y}) + h_{11mn}(x_1, \mathbf{y}) K^1_{m,ny}(\mathbf{y}) \rangle ,$$

$$N^* = \langle \sigma^*_{11}(x_1, \mathbf{y}) \rangle ,$$

$$R_\beta(\sigma) = \langle h_{\beta 111}(x_1, \mathbf{y}) + h_{\beta 1mn}(x_1, \mathbf{y}) K^1_{m,ny}(\mathbf{y}) \rangle .$$

Here, N^* is the initial axial stress in the string, $A(\sigma)$ is the homogenized axial stiffness of the string, $u^{(0)}_1$ is the axial displacement, and $u^{(0)}_\alpha$ $(\alpha = 2,3)$ are the normal deflections of the string. The constant A, generally, depends on the initial stresses in the plane of the string. This dependence is analogous to that found by Kolpakov (1990, 1992) and described briefly in Sect. 1.3.

The boundary conditions

The boundary conditions for axial displacement $u^{(0)}_1$ and deflections $u^{(0)}_\alpha$ $(\alpha = 2,3)$, obtained by substituting expansion (3.7.5) in the original boundary conditions on $S_u = \{\mathbf{x} \in G_\varepsilon : x_1 = \pm a\}$ [see (3.7.1)], have the form,

$$u^{(0)}_1(-a) = u^{(0)}_1(a) = 0 , \tag{3.7.31}$$

$$u_\alpha^{(0)}(-a) = u_\alpha^{(0)}(a) = 0 \ (\alpha = 2, 3) . \tag{3.7.32}$$

When $\langle f_1 \rangle = 0$ and $\langle g_1 \rangle_S = 0$, the equilibrium equations (3.7.24) with constitutive equations (3.7.29) and boundary conditions (3.7.31), have the unique solution $u_\alpha^{(0)}(x_1) \, (\alpha = 2, 3)$ under the condition that the initial stresses do not cause stability loss in the string as a 1-D body. This condition is nearly always satisfied in practice since the initial stresses are small as compared with the elasticity constants. Then, (3.7.30) takes the form,

$$N_\beta = N^* u_{\beta,1x}^{(0)}(x_1) , \tag{3.7.33}$$

where N_β satisfies the equilibrium equations [see (3.7.25)],

$$N_{\beta,1x} + \langle f_\beta \rangle + \langle g_\beta \rangle_S = 0 . \tag{3.7.34}$$

Equations (3.7.33), (3.7.34) can be rearranged into a 1-D model of a string,

$$(N^* u_{\beta,1x}^{(0)})_{,1x} + \langle f_\beta \rangle + \langle g_\beta \rangle_S = 0 , \tag{3.7.35}$$

with the boundary conditions (3.7.32)

The actual stresses in a string

The total stresses Σ_{ij} in a string are the sum of initial stresses $\overset{*}{\sigma}_{ij}$ and additional stresses σ_{ij}:

$$\Sigma_{ij}(x_1, \mathbf{y}) = \overset{*}{\sigma}_{ij}(x_1, \mathbf{y}) + \sigma_{ij}(x_1, \mathbf{y}) . \tag{3.7.36}$$

Saving the leading term [see (3.7.6), (3.7.20)], we have the following approximation for the additional stresses:

$$\sigma_{ij}(x_1, \mathbf{y}) \approx [-\overset{*}{\sigma}_{j\alpha}(x_1, \mathbf{y})\delta_{i1} + \overset{*}{\sigma}_{j1}(x_1, \mathbf{y})\delta_{i\alpha}] u_{\alpha,1x}^{(0)}(x_1) . \tag{3.7.37}$$

It is seen that the elastic constants do not influence additional stresses σ_{ij} (3.7.37). It is in agreement with classical theory. This elastic constants influence initial stresses $\overset{*}{\sigma}_{ij}$. As was noted above, it is impossible to introduce initial stresses in an elastic body when the elastic constants are neglected.

Comparison with classical theory. Strings of uniform and coaxial structure

Consider a uniform string made of a homogeneous material. In this case, periodicity cell $Y = [0, T] \times R$ (see Fig. 3.6), where R is the cross section of the string:

$$T^{-1} \int_Y \mathrm{d}\mathbf{y} = \int_R \mathrm{d}y_2 \mathrm{d}y_3 .$$

As a result, we arrive at the classical definition, introducing the forces in a uniform string by integrating over the cross section of the string. Then, the asymptotic model developed is in agreement with the classical theory for a string. Similar results can be obtained for strings of coaxial structure (Fig. 3.6).

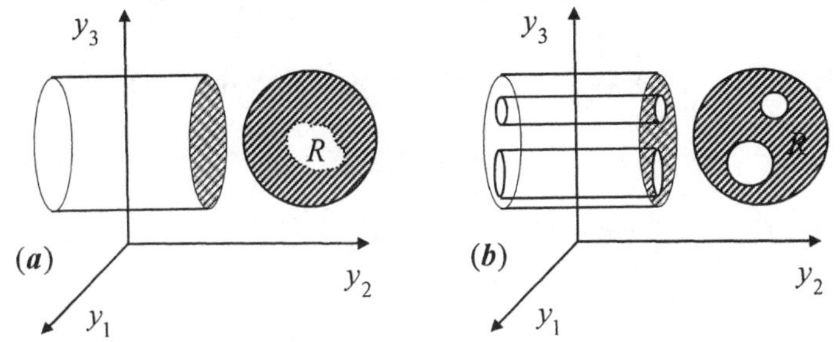

Fig. 3.6. Periodicity cell and cross section of a uniform string (***a***); a string of coaxial structure (***b***)

If the axial force $\langle f_1 \rangle + \langle g_1 \rangle_S$ is not equal to zero, the displacement $u_1^{(0)}(x_1)$ may not be equal to zero, and we have a coupled model. However, we do not see a coupling effect in uniform strings. We give an explanation of this fact. Consider a uniform string of a coaxial structure made of homogeneous isotropic materials. In the case under consideration, the solution of cellular problem (3.7.13) can be found in the form $X_1 = 0$, $X_\beta = X_\beta(x_2, x_3)$ ($\beta = 2,3$). For this function,

$$[h_{ijmn}(x_1, \mathbf{y}) K^p_{m,ny} + h_{ijp1}(x_1, \mathbf{y})]_{,jy} \tag{3.7.38}$$

$$= [h_{i\alpha\gamma\beta}(x_1, \mathbf{y}) K^p_{m,ny} + h_{i\alpha p1}(x_1, \mathbf{y})]_{,\alpha y} \quad \text{in } R ,$$

$$[h_{ijmn}(x_1, \mathbf{y}) K^p_{m,ny} + h_{ijp1}(x_1, \mathbf{y})] n_j$$

$$= [h_{i\alpha\gamma\beta}(x_1, \mathbf{y})K^p_{m,ny} + h_{i\alpha p1}(x_1, \mathbf{y})]n_\alpha \quad \text{on } \partial R\,.$$

Here α, γ, β only take values 2, 3.

For $i = 1$, the expression $h_{i\alpha\gamma\beta}(x_1, y_2, y_3)K^1_{\gamma,\beta y} + h_{i\alpha 11}(x_1, y_2, y_3)$ from (3.7.38) takes the form, $h_{1\alpha\gamma\beta}(x_1, y_2, y_3)K^1_{\gamma,\beta y} + h_{1\alpha 11}(x_1, y_2, y_3)$. This expression is equal to zero because

$$h_{1\alpha 11}(x_1, y_2, y_3) = c_{1\alpha 11}(y_2, y_3) + \overset{*}{\sigma}_{\alpha 1}(x_1, y_2, y_3) = 0$$

due to $c_{1\alpha 11} = 0$ for an isotropic material and $\overset{*}{\sigma}_{\alpha 1} = 0$ in a uniform homogeneous beam (the equality $\overset{*}{\sigma}_{\alpha 1} = 0$ is known from Kolpakov, 1994) and

$$h_{1\alpha\gamma\beta}(x_1, y_2, y_3) = c_{1\alpha\gamma\beta}(y_2, y_3) + \overset{*}{\sigma}_{\alpha\beta}(x_1, y_2, y_3)\delta_{1\gamma} = 0$$

due to $c_{1\alpha\gamma\beta} = 0$ for an isotropic material and $\delta_{1\gamma} = 0$. For $i = 2, 3$, we obtain two equations with respect to the two functions $K^1_\beta = K^1_\beta(y_2, y_3)$ ($\beta = 2, 3$).

Thus,

$$R_\beta(\sigma) = \langle h_{\beta 111}(x_1, y_2, y_3) + h_{\beta 1mn}(x_1, y_2, y_3)X^1_{m,ny}(y_2, y_3)\rangle$$

$$= \langle h_{\beta 111}(x_1, y_2, y_3) + h_{\beta 1\gamma\alpha}(x_1, y_2, y_3)X^1_{\gamma,\alpha y}(y_2, y_3)\rangle = 0$$

because

$$h_{\beta 111}(y_2, y_3) = c_{\beta 111}(y_2, y_3) + \overset{*}{\sigma}_{11}(x_1, y_2, y_3)\delta_{\beta 1} = 0$$

and

$$h_{\beta 111}(x_1, y_2, y_3) = c_{\beta 111}(y_2, y_3) + \overset{*}{\sigma}_{1\alpha}(x_1, y_2, y_3)\delta_{\beta\gamma}\rangle\rangle = 0$$

due to the equalities $c_{\beta 111} = 0$, $\delta_{\beta 1} = 0$, $c_{\beta 111} = 0$, and $\overset{*}{\sigma}_{1\alpha} = 0$.

The dynamic problem

We make transition from a 3-D dynamic elasticity problem to a 1-D string model. We propose that the dynamic effects are not "fast", namely, $\partial^2 u_i^\varepsilon / \partial t^2$ has the order of unity. In particular, if we consider an eigenvibration of the string, it means that we consider the low-frequency vibrational mode.

The dynamic equations for a body with initial stresses have the form (see, e.g., Washizu, 1982)

$$\frac{\partial \sigma_{ij}}{\partial x_j}(\mathbf{x}, \mathbf{x}/\varepsilon) + \varepsilon^{-2} f_i(\mathbf{x}, \mathbf{x}/\varepsilon, t) \tag{3.7.39}$$

$$= \varepsilon^{-2} \rho(\mathbf{x}/\varepsilon) \frac{\partial^2 u_i^\varepsilon}{\partial t^2} \quad \text{in } G_\varepsilon,$$

$$\sigma_{ij} n_j = \varepsilon^{-1} g_i(\mathbf{x}, \mathbf{x}/\varepsilon, t) \quad \text{on } S_\varepsilon,$$

$$\mathbf{u}^\varepsilon(\mathbf{x}) = 0 \quad \text{on } S_u = \{ \mathbf{x} \in G_\varepsilon : x_1 = \pm a \}.$$

By substituting expansions (3.7.5) and (3.7.6) in (3.7.39) and taking into account (3.7.4), we find that

$$\sum_{p=-2}^{\infty} \varepsilon^p (\varepsilon^{-1} \sigma_{ij,jy}^{(p)} + \sigma_{i1,1x}^{(p)}) + \varepsilon^{-2} f_i(\mathbf{x}, \mathbf{x}/\varepsilon, t)$$

$$= \sum_{k=0}^{\infty} \varepsilon^{k-2} \rho(\mathbf{y}) \frac{\partial^2 u_i^{(k)}}{\partial t^2} \quad \text{in } Y,$$

$$\sum_{p=-2}^{\infty} \varepsilon^p \sigma_{ij}^{(p)} n_j = \varepsilon^{-1} g_i(\mathbf{x}, \mathbf{x}/\varepsilon, t) \quad \text{on } S.$$

Equating the terms accompanying the same powers of ε, we obtain the following equations:

$$\sigma_{ij,jy}^{(p)} + \sigma_{ij,1x}^{(p-1)} = \begin{cases} 0 & \text{if } p = -2, \\ -f_i + \rho(\mathbf{y}) \dfrac{\partial^2 u_i^{(0)}}{\partial t^2} & \text{if } p = -1, \quad \text{in } Y; \\ \rho(\mathbf{y}) \dfrac{\partial^2 u_i^{(p+2)}}{\partial t^2} & \text{if } p > -1, \end{cases} \tag{3.7.40}$$

$$\sigma_{ij}^{(p)} n_j = \begin{cases} 0 & \text{if } p \neq -1, \\ g_i & \text{if } p = -1, \end{cases} \quad \text{on } S.$$

As above, we determine that $\sigma_{ij}^{(-3)} = 0$.

We derive the overall dynamic equations. The first equation in (3.7.39) can be written in the following form ($\mathbf{v}$ is the trial function):

$$\int_{G_\varepsilon} \sigma_{ij} \frac{\partial v_i}{\partial x_j} \mathbf{dx} = \varepsilon^{-1} \int_{S_\varepsilon} \mathbf{gvdx} + \varepsilon^{-2} \int_{G_\varepsilon} \mathbf{fvdx} \tag{3.7.41}$$

$$- \varepsilon^{-2} \int_{G_\varepsilon} \rho(\mathbf{x}/\varepsilon) \frac{\partial^2 u_i^\varepsilon}{\partial t^2} \mathbf{vdx} \quad \text{for any } \mathbf{v} \in U_\varepsilon .$$

Substituting (3.7.6) in (3.7.41) and taking into account differentiating rule (3.7.4), we obtain

$$\sum_{p=-2}^{\infty} \int_{G_\varepsilon} \varepsilon^p (\varepsilon^{-1} \sigma_{ij}^{(p)} v_{i,jy} + \sigma_{i1}^{(p)} v_{i,1x}) \mathbf{dx} \tag{3.7.42}$$

$$= \varepsilon^{-1} \int_{S_\varepsilon} \mathbf{gvdx} + \varepsilon^{-2} \int_{G_\varepsilon} \mathbf{fvdx} - \sum_{k=0}^{\infty} \varepsilon^{-2+k} \int_{G_\varepsilon} \rho(\mathbf{x}/\varepsilon) \frac{\partial^2 u_i^{(k)}}{\partial t^2} \mathbf{vdx}$$

$$\text{for any } \mathbf{v} \in U_\varepsilon .$$

We take the trial function in (3.7.42) in the form, $\mathbf{v} = \mathbf{w}(x_1) \in C^1([-a,a])$. Taking into account Proposition 3.2, we obtain the following equality from (3.7.42):

$$\sum_{p=-2}^{\infty} \int_{-a}^{a} \langle \sigma_{i1}^{(p)} \rangle w_{i,1x} \mathbf{dx}_1 = \tag{3.7.43}$$

$$\int_{-a}^{a} \langle \mathbf{g} \rangle_S \mathbf{wdx}_1 + \int_{-a}^{a} \langle \mathbf{f} \rangle \mathbf{wdx}_1$$

$$-\sum_{k=0}^{\infty} \varepsilon^{-2+k} \int_{-a}^{a} \langle \rho \rangle \frac{\partial^2 u_i^{(k)}}{\partial t^2} \mathbf{w} dx_1$$

for any $\mathbf{w}(x_1) \in C^1([-a, a])$.

Integrating (3.7.43) by parts and equating the expressions corresponding to ε^0 [it is the leading term in (3.7.43)], we obtain the following 1-D dynamic equations:

$$N_{1,1x} + \langle f_1 \rangle + \langle g_1 \rangle_S = \langle \rho \rangle \frac{\partial^2 u_i^{(0)}}{\partial t^2}, \tag{3.7.44}$$

$$N_{\alpha,1x} + \langle f_\alpha \rangle + \langle g_\alpha \rangle_S = \langle \rho \rangle \frac{\partial^2 u_\alpha^{(0)}}{\partial t^2}. \tag{3.7.45}$$

Constitutive equations for a string considered as a 1-D structure

The static 1-D constitutive equation for a string was derived from (3.7.8, $p = -2$) and (3.7.9, $p = -2$). Equation (3.7.8, $p = -2$) does not depend on the static or dynamic problem considered. Equations (3.7.40) [the dynamic analog of static equations (3.7.9)] depend on the problem we are considering. Nevertheless, equation (3.7.40, $p = -2$) does not depend on this problem. Thus, the 1-D constitutive equation is the same for the static and dynamic cases. These are

$$N_1 = A(\sigma) u_{1,1x}^{(0)}(x_1), \tag{3.7.46}$$

$$N_\beta = N^* u_{\beta,1x}^{(0)}(x_1) + R_\beta u_{1,1x}^{(0)}(x_1), \tag{3.7.47}$$

where N_β satisfies the equilibrium equations (3.7.44), (3.7.45).

Two Special Cases

The case $\langle f_1 \rangle + \langle g_1 \rangle_S = 0$

It follows from (3.7.44), (3.7.46) that the axial the axial displacement $u_1^{(0)}(x_1, t)$ is determined independently on the transversal displacements. If $\langle f_1 \rangle + \langle g_1 \rangle_S = 0$ and the initial conditions uniform, then $u_1^{(0)}(x_1, t)$.

Cylindrical uniform string

For a cylindrical uniform string in any case, $R_\beta = 0$.

Thus, for both the cases considered here, (3.7.47) takes the form $N_\beta = N^* u^{(0)}_{\beta,1x}(x_1)$ and we arrive at classical model, involving only the transversal displacements of string,

$$(N^* u^{(0)}_{\beta,1x})_{,1x} + \langle f_\beta \rangle + \langle g_\beta \rangle_S = \langle \rho \rangle \frac{\partial^2 u^{(0)}_\beta}{\partial t^2}, \qquad (3.7.48)$$

with the boundary conditions (3.7.32).

An example

Consider a string, which consists of a high-modulus cylindrical core and a soft cover; see Fig. 3.7.

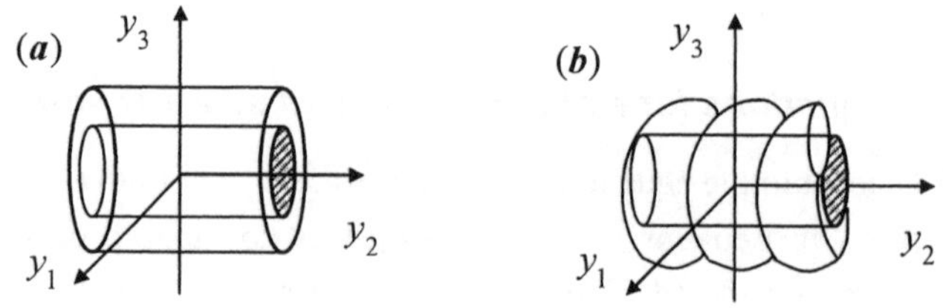

Fig. 3.7. The "high-modulus core, soft cover" strings: (***a***) the real and (***b***) the constructive soft covers

We consider the initial stresses in such strings. When we apply an axial stress to the string, the corresponding local stresses are concentrated in the high-modulus core, and the local stresses in the soft cover are small and may be neglected. Thus, if the strings have different soft covers and the same high-modulus core, the axial forces N^* are the same for these strings. Then, 1-D models of the strings are equivalent to one another while we consider the static problem.

At the same time, the soft cover can have a density compared with the density of the core. Then, the cover can significantly influence the average density $\langle \rho \rangle$ in the 1-D model of the string (3.7.39).

These predictions of asymptotic theory are in full agreement with the fact we see in practice. Fig. 3.7(***b***) shows a version of the "high-modulus core, soft cover" - the constructive soft cover. In this case, the cover works like a spring and has a small stiffness in the direction of the Ox_2 axis. This type of string is widely used in musical instruments.

3.8 Beams with no Initial Stresses. Computing of Resultant Initial Stresses and Moments

The theory of beams with no initial stresses is important in practical applications. It is also very important for the theory of stressed beams. In the sections above, we introduced the resultant initial axial stress and the resultant initial moments by integrating local stresses and local moments over the periodicity cell. These definitions create no problems for the theoretical analysis of stressed beams, but they are not suitable for practical computations. In this section, we demonstrate that the resultant axial stress and moments, defined as above, are equal to the resultant axial stress and moments computed using a 1-D homogenization model of a beam. It means that the inhomogeneous beam (beam of complex structure) can be analyzed within the framework of 1-D models. This proposition cannot be extended to the problem of computing of homogenized stiffnesses, which is essentially 3-D problem.

The homogenized model for a beam with no initial stresses

The asymptotic theories for a beam were developed by Trabucho and Viaño (1987) and Tutek and Aganovich (1987), see also extensive references in the book by Trabucho and Viaño (1996). These works were devoted to the asymptotic analysis of the 3-D elasticity problem in a cylindrical domain of small diameter. In other words, these works were devoted to the analysis and justification of classical beam models. Trabucho and Viaño (1987, 1996) consider some problems for a beam with a slow varying cross section (in our terms, for cross section depending on "slow" variable $\mathbf{x}$ but not "fast" variable $\mathbf{x}/\varepsilon$. Trabucho and Viaño (1996) consider also the homogenization problem corresponding to fibers/holes parallel to the axis of the beam (the so-called coaxial structure). The restriction on the domain - a cylinder only, was essential and related to the geometrical rescaling of the domain. We mention the paper by Kozlova (1989) devoted to the homogenization problem for an inhomogeneous beam of coaxial structure. The formal homogenization procedure for beams of general periodic structure was given by Kolpakov (1991).

A homogenized model of a beam with no initial stresses can be obtained from the model presented in Sect. 3.2. It is sufficient to put $\sigma_{ij}^{*(p)}(x_1,\mathbf{y})\,(p=-3,-2)$ in (3.2.3) and the following formulas. We present here the more important formulas, which can be obtained in this way.

The following approximation for local stresses is valid [see (3.2.44)]:

$$\sigma_{ij}^{(-3)} \approx [c_{ijmn}(\mathbf{y})X_{m,ny}^{01}(\mathbf{y}) - c_{ij11}(\mathbf{y})]V_{1,1x}(x_1) \qquad (3.8.1)$$

$$+[c_{ijmn}(\mathbf{y})X^{1\alpha}_{m,ny}(\mathbf{y})+c_{ij11}(\mathbf{y})y_\alpha]u^{(0)}_{\alpha,1x1x}(x_1)$$

$$+[c_{ijmn}(\mathbf{y})X^{3}_{m,ny}(\mathbf{y})+c_{ij\gamma 1}(\mathbf{y})s_\gamma y_\Gamma]\varphi_{,1x}(x_1)\,.$$

Here, V_1 ,$u^{(0)}_\alpha$,φ are the axial displacement, the normal deflections, and the torsion angle, respectively. They are determined from the solution of the homogenized problem for a beam with no initial stresses. The homogenized problem consists of the following constitutive equations:

$$N^{(-3)}_{11} = AV_{1,1x} + A^1_\alpha u^{(0)}_{\alpha,1} + b\varphi_{,1x}\,, \tag{3.8.2}$$

$$M^{(-3)}_{1\beta} = {}^1A_\beta V_{1,1x} + A^2_{\beta\alpha} u^{(0)}_{\alpha,1x1x} + B_\beta \varphi_{,1x}\,, \tag{3.8.3}$$

$$M = A^1 V_{1,1x} + A^2_\alpha u^{(0)}_{\alpha,1x1x} + B\varphi_{,1x}\,, \tag{3.8.4}$$

and equations of equilibrium (3.2.49)–(3.2.52).

Here, $N^{(-3)}_{11}$ is the resultant axial force, $M^{(-3)}_{1\beta} = \langle \sigma^{(-3)}_{11} y_\beta \rangle$ are the bending moments, and $M = M^{(-3)}_{32} - M^{(-3)}_{23} = \langle \sigma^{(-3)}_{31} y_2 - \sigma^{(-3)}_{21} y_3 \rangle$ is the torsional moment.

The homogenized stiffnesses (the coefficients of the homogenized constitutive equations) are given by (3.2.48). The homogenized stiffnesses can be computed in accordance with (3.2.48) only after cellular problems (3.2.29), (3.2.41), and (3.2.42) are solved.

Rods

If we neglect bending and torsion, the homogenized model obtained describes the homogenized rod corresponding to the initial inhomogeneous rod. Taking into account the practical importance of rods, we write here the homogenized model for it. The model is as follows.

The equilibrium equation,

$$N^{(-3)}_{11,1x} = 0\,.$$

The homogenized constitutive equation

$$N^{(-3)}_{11} = AV_{1,1x}\,,$$

where A is the homogenized stiffness of rod

$$A = \langle c_{1111}(\mathbf{y}) + c_{11mn}(\mathbf{y}) X^{01}_{m,ny}(\mathbf{y}) \rangle ,$$

and $\mathbf{X}^{01}(\mathbf{y})$ is the solution of the cellular problem,

$$\begin{cases} [c_{ijmn}(\mathbf{y}) X^{0p}_{m,ny} + c_{ijp1}(\mathbf{y})]_{,jy} = 0 \quad \text{in } Y, \\ [c_{ijmn}(\mathbf{y}) X^{0p}_{m,ny} + c_{ijp1}(\mathbf{y})] n_j = 0 \quad \text{on } S, \\ \mathbf{X}^{0p}(\mathbf{y}) \text{ is periodic in } y_1 \text{ with period } T. \end{cases}$$

We consider equality (3.5.14) for the case $\sigma^*_{ij} = 0$. The quantity $E(0, \mathbf{u})$ is the elastic energy of a beam (rod), considered as a 3-D elastic body. Thus,

$$E(\sigma, \mathbf{u}) \approx E(u_2^{(0)}, u_3^{(0)}, V_1, \phi) .$$

where $E(u_2^{(0)}, u_3^{(0)}, V_1, \phi)$ is the energy computed using 1-D model.

Computing resultant initial stresses and moments

1-D models of beams with initial stresses and 1-D string models involve the resultant axial force N^*_{11} and bending and torsional moments $M^*_{1\beta}$, M^* introduced by averaging local stresses $\sigma^*_{11}(x_1, \mathbf{y})$ and local moments $\sigma^*_{i1}(x_1, \mathbf{y}) y_\beta$ over 3-D periodicity cell Y:

$$N^*_{11} = T^{-1} \int_Y \sigma^*_{11}(x_1, \mathbf{y}) d\mathbf{y} , \tag{3.8.5}$$

$$M^*_{1\beta} = T^{-1} \int_Y \sigma^*_{11}(x_1, \mathbf{y}) y_\beta d\mathbf{y} ,$$

$$M^* = T^{-1} \int_Y [\sigma^*_{31}(x_1, \mathbf{y}) y_2 - \sigma^*_{21}(x_1, \mathbf{y}) y_3] d\mathbf{y} .$$

where $\sigma^*_{ij}(x_1, \mathbf{y})$ is defined by solving the 3-D elasticity problem for a body of small diameter. This method is not effective because we have to solve the 3-D elasticity problem.

We prove that the resultant initial stresses and moments problem can be determined by solving a 1-D problem with no initial stresses. This proof will give completeness to the developed theory of 1-D beams and strings of complex structure. After we prove that the resultant initial stresses and moments problem can be determined by solving a 1-D problem, we will have a theory involving only 1-D models.

We write down (3.8.1) for the original state of an inhomogeneous beam in the following form:

$$\sigma^*_{ij} = [c_{ijmn}(\mathbf{y})X^{01}_{m,ny}(\mathbf{y}) - c_{ij11}(\mathbf{y})]V_{1,1x}(x_1) \tag{3.8.6}$$

$$+[c_{ijmn}(\mathbf{y})X^{1\alpha}_{m,ny}(\mathbf{y}) + c_{ij11}(\mathbf{y})y_\alpha]u^{(0)}_{\alpha,1x1x}(x_1)$$

$$+[c_{ijmn}(\mathbf{y})X^{3}_{m,ny}(\mathbf{y}) + c_{ij\gamma 1}(\mathbf{y})s_\gamma y_\Gamma]\varphi_{,1x}(x_1)\,.$$

Averaging (3.8.6) with $ij = 11$ over periodicity cell Y, we obtain

$$N^*_{11} = AV_{1,1x} + A^1_\alpha u^{(0)}_{\alpha,1} + b\varphi_{,1x}\,.$$

Multiplying (3.8.6) with $j = 1$ by y_β and averaging over periodicity cell Y, we obtain

$$M^*_{1\beta} = {}^1A_\beta V_{1,1x} + A^2_{\beta\alpha}u^{(0)}_{\alpha,1x1x} + B_\beta\varphi_{,1x}\,, \tag{3.8.7}$$

$$M^* = A^1V_{1,1x} + A^2_\alpha u^{(0)}_{\alpha,1x1x} + B\varphi_{,1x}\,. \tag{3.8.8}$$

The right-hand parts of (3.8.7) and (3.8.8) involve only solutions of a 1-D problem.

Both methods described above (the averaging of local stresses and moments and the determination of the resultant stresses and moments from the solution of a 1-D beam or rod problem) incorporate the structure of a beam. The first method incorporates the structure of a beam in an explicit form. The second method incorporates the structure of a beam through the homogenized stiffnesses of the beam.

4 Calculation and Estimation of Homogenized Stiffnesses of Plate-Like and Beam-Like Composite Structures

As follows from Chap. 2 and 3, the analysis of plates and beams of complex structure is reduced to computing the homogenized stiffnesses and the resultant initial stresses and moments. As demonstrated in Sects. 2.7 and 3.6, the resultant initial stresses and moments can be computed from the solutions of corresponding homogenized problems. This means that one can analyze plate-like and beam-like structures within the framework of the corresponding 2-D and 1-D models. There exists only one problem, which cannot be solved within the framework of 2-D and 1-D models, the problem of computing of effective stiffnesses. To solve this problem, it is necessary to solve 3-D cellular problems. Thus, computing the homogenized stiffnesses of plates and beams with no initial stresses is the principal problem for the theory of stressed plates and beams. For this reason, the author presents methods for computing and estimating the homogenized stiffnesses of plates and beams of complex structure.

The following methods are discussed: variational principles for computing stiffnesses of plates and beams of complex structure; modification of the homogenization method for lattice structures; and solution of cellular problems of the elasticity theory with commercial software.

4.1 Variational Principles for Stiffnesses of Nonhomogeneous Plates

Homogenized constants (also called effective, overall, or macroscopic) are related to variational principles. The first variational principles and two-sided estimates for homogenized constants of 3-D composites were obtained by Voight (1928), Reuss (1929), and Hill (1952). After them, variational principles and two-sided estimates were investigated in numerous works: see Hashin and Shtrikman (1963), Yeh (1970), Beran and Molyneux (1966), and Willis (1981). They have been derived elasticity theory methods (see, e.g., Sendeckyj, 1974, Nemat–Nasser and Hori, 1993) and the homogenization method (see, e.g., Francfort and Murat, 1986, Milton and Kohn, 1988). In the works mentioned, variational principles and estimates for homogenized constants of 3-D (solid-like) composites were considered. Variational principles and estimates for homogenized constants (stiffnesses) of composite beams and plates have been derived by Kolpakov (1998b, 1999).

In this section, variational principles are derived for a nonhomogeneous plate of arbitrary structure. Lagrange and Castigliano-type variational principles for an inhomogeneous plate of periodic structure are derived using the homogenization method presented in Chap. 2. The principles obtained provide, in particular,

Voight–Reuss-like bounds for stiffnesses.

Rigorous variational principles for homogenized stiffnesses must be derived directly from a 3-D elasticity problem without any additional hypothesis. For a monolithic body, this creates no problem. For a plate, one must, first, relate the 3-D elasticity problem for a thin body to 2-D plate strain characteristics (in-plane strains and curvatures). This can be done using the homogenization method for plates presented in Chap. 2.

Statement of the problem

Consider a thin elastic body formed by periodic repetition of periodicity cell εY of small dimension ε in the Ox_1x_2 plane. A detailed description of the body under consideration was given in Sect. 2.2 (see Fig. 2.2). The tensor of local elastic constants of the body considered has the form, $c_{ijmn}(\mathbf{x}/\varepsilon)$, where $c_{ijmn}(\mathbf{y})$ are periodic functions with respect to y_1, y_2 with periodicity cell T. Periodicity cell Y in "fast" variables $\mathbf{y} = \mathbf{x}/\varepsilon$ is shown in Fig. 4.1; T is the projection of Y on the Oy_1y_2 plane. It is assumed that the local elastic constants satisfy conditions C from Sect. 1.2.

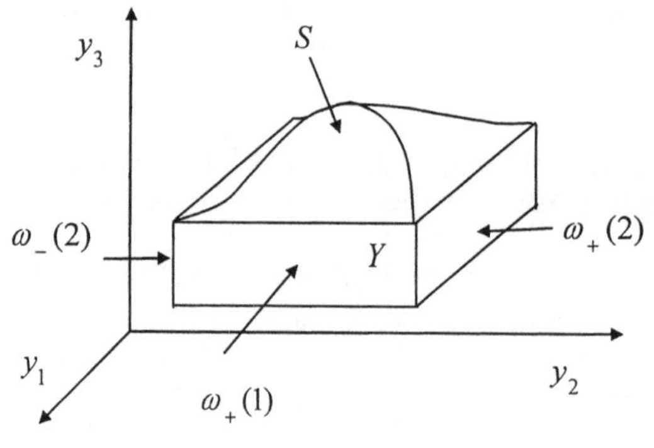

Fig. 4.1. A periodicity cell of a plate in "fast" variables

The plate stiffnesses $A^{\nu+\mu}_{\alpha\beta\gamma\delta}$ are calculated as follows. In the first step, the cellular problem is solved. In the case under consideration, the cellular problem has the form (see Sect. 2.2),

$$\begin{cases} [c_{ijmn}(\mathbf{y})X^{\nu\alpha\beta}_{m,n} + (-1)^{\nu} y_3^{\nu} c_{ij\beta\alpha}(\mathbf{y})]_{,j} = 0 \quad \text{in } Y, \\ [c_{ijmn}(\mathbf{y})X^{\nu\alpha\beta}_{m,n} + (-1)^{\nu} y_3^{\nu} c_{ij\beta\alpha}(\mathbf{y})]n_j = 0 \quad \text{on } S, \\ \mathbf{X}^{\nu\alpha\beta}(\mathbf{y}) \text{ is periodic in } y_1, y_2 \text{ with periodicity cell } T \end{cases} \tag{4.1.1}$$

$$(\nu = 0, 1),$$

where **n** is the normal to the free surface S of domain Y (see Fig.4.1).

In this chapter, we consider only cellular problems, which are all written in fast variables $\mathbf{y}$. Subscript $,j$ means $\partial / \partial y_j$. The Latin indexes take values 1, 2, 3; Greek indexes $\alpha, \beta, \gamma, \delta$ take values 1, 2, and superscripts ν, μ take values 0, 1.

After solving problem (4.1.1), the stiffnesses are calculated in accordance with the formula,

$$A^{\nu+\mu}_{\gamma\delta\alpha\beta} = \langle [(-1)^{\nu} c_{\gamma\delta\alpha\beta}(\mathbf{y}) y_3^{\nu} + c_{\gamma\delta mn}(\mathbf{y}) X^{\nu\alpha\beta}_{m,n}(\mathbf{y})](-1)^{\mu} y_3^{\mu} \rangle, \tag{4.1.2}$$

where $\langle \circ \rangle = (mes T)^{-1} \int_Y \circ \, d\mathbf{y}$.

Formula (4.1.2) gives the bending stiffnesses $A^2_{\alpha\beta\alpha\beta}$ when $\mu = \nu = 1$, the coupling (also called out-of-plane) stiffnesses $A^1_{\alpha\beta\alpha\beta}$ when $\mu + \nu = 1$, and the in-plane stiffnesses $A^0_{\alpha\beta\alpha\beta}$ when $\mu = \nu = 0$.

For a plate with planar lateral surfaces in the paper by Caillerie (1984), the following formula was presented:

$$A^{\nu+\mu}_{\gamma\delta\alpha\beta} = \langle c_{ijkl}(\mathbf{y})[X^{\nu\gamma\delta}_{k,l}(\mathbf{y}) + (-1)^{\nu} y_3^{\nu} \delta_{k\gamma}\delta_{l\delta}][X^{\nu\alpha\beta}_{i,j}(\mathbf{y}) + (-1)^{\mu} y_3^{\mu} \delta_{i\alpha}\delta_{j\beta}] \rangle, \tag{4.1.3}$$

giving stiffnesses in the form of a quadratic functional.

We prove that (4.1.3) remains valid for plates with nonplanar lateral surfaces (ribbed, wavy, lattice, and so on). For that, let us derive (4.1.3) directly from (4.1.1), (4.1.2). Multiplying the first equation in (4.1.1) by $X^{\gamma\delta\mu}_l$, integrating by parts, and taking into account the boundary and periodicity conditions from (4.1.1), we obtain the following equality:

$$0 = \langle c_{ijkl}(\mathbf{y})[X^{\nu\alpha\beta}_{k,l} + (-1)^{\nu} y_3 \delta_{k\alpha}\delta_{l\beta}] X^{\mu\gamma\delta}_{k,l} \rangle . \tag{4.1.4}$$

Adding (4.1.4) and (4.1.2), we obtain (4.1.3).

As follows from (4.1.3), the stiffnesses are symmetrical in indexes $\alpha, \beta, \gamma, \delta$: $A^{\nu+\mu}_{\gamma\delta\alpha\beta} = A^{\nu+\mu}_{\alpha\beta\gamma\delta}$.

Variational principles and estimates for bending stiffnesses

In this section, the variational principles for bending stiffnesses $A^2_{\alpha\beta\alpha\beta}$ are derived. The stiffnesses are expressed through the extreme values of Lagrange and Castigliano functionals defined on periodicity cell Y of plate.

Bending stiffnesses $A^2_{\alpha\beta\alpha\beta}$ are given by (4.1.3) with $\alpha\beta = \gamma\delta$, $\mu = \nu = 1$. Let us establish the relation between the Lagrange and Castigliano functionals for problem (4.1.1) and bending stiffnesses.

The Lagrange functional $J_u(\mathbf{u})$ for cellular problem (4.1.1) with $\alpha\beta = \gamma\delta$, $\mu = \nu = 1$ is the following:

$$J_u(\mathbf{u}) = (1/2)\langle 2y_3^\nu c_{ij\alpha\beta}(\mathbf{y})u_{i,j} - c_{ijkl}(\mathbf{y})u_{i,j}u_{k,l}\rangle \ . \tag{4.1.5}$$

It is considered on the set of virtual displacements,

$$V = \{\mathbf{u} \in \{H^1(Y)\}^3 : \mathbf{u}(\mathbf{y}) \text{ is periodic in } y_1, y_2 \text{ with periodicity cell } T\}. \tag{4.1.6}$$

Let us introduce (in formally, for the moment) the Castigliano functional as follows:

$$J_\sigma(\sigma) = (1/2)\langle c^{-1}_{ijkl}(\mathbf{y})\sigma_{ij}\sigma_{kl} + 2\sigma_{\alpha\beta}y_3 + y_3^2 c_{\alpha\beta\alpha\beta}(\mathbf{y})\rangle \ . \tag{4.1.7}$$

(c^{-1}_{ijkl} means the tensor inverse with respect to tensor c_{ijkl}) considered on the set of admissible stresses,

$$\Sigma = \{\sigma_{ij}(\mathbf{y}) \in \{L_2(Y)\}^6 : \sigma_{ij,j} = 0 \ \text{ in Y}, \ \sigma_{ij}n_j = 0 \ \text{ on S}, \tag{4.1.8}$$

$$\sigma_{ij}n_j \ \text{ is periodic in } y_1, y_2 \text{ with periodicity cell } T\} \ .$$

We establish the relation between Lagrange and Castigliano functionals.

Proposition 4.1. *Under conditions C* (*Sect.* 1.2), *the following equalities hold*:

$$\max_{\mathbf{u}\in V} J_u(\mathbf{u}) = \min_{\sigma\in\Sigma} J_\sigma(\sigma) = J_u(\mathbf{X}^{1\alpha\beta}), \tag{4.1.9}$$

where $\mathbf{X}^{1\alpha\beta}$ *is the solution of problem* (4.1.1) *with* $\nu = 1$.

Proof. Under conditions C, the functional $J_u(\mathbf{u})$ is strongly convex (see

Ekeland and Temam, 1976) on the set $\{\ \mathbf{u} \in V : \langle \mathbf{u} \rangle = 0\ \}$. Then, the problem

$$J_u(\mathbf{u}) \to \max,\ \mathbf{u} \in V, \tag{4.1.10}$$

has a unique solution belonging to the set $\{\mathbf{u} \in V : \langle \mathbf{u} \rangle = 0\}$, and problem (4.1.1) is Euler's equation for problem (4.1.10). Then, $\mathbf{X}^{1\alpha\beta}$ is the solution of problem (4.1.10) and $\max\limits_{\mathbf{u} \in V} J_u(\mathbf{u}) = J_u(\mathbf{X}^{1\alpha\beta})$.

The Castigliano functional (4.1.7) under condition C, Sect. 1.2, is strongly convex (Ekeland and Temam, 1976). Then, the problem,

$$J_u(\mathbf{u}) \to \max,\ \mathbf{u} \in V, \tag{4.1.11}$$

has a unique solution, which satisfies the following equation:

$$\int_Y (c_{ijkl}(\mathbf{y})\sigma_{kl} + c_{ij\alpha\beta}(\mathbf{y}))\eta_{ij}(\mathbf{y})d\mathbf{y} = 0 \quad \text{for any } \eta_{ij} \in \Sigma. \tag{4.1.12}$$

Let us consider the stresses corresponding to the solution of the problem (4.1.1)

$$\sigma_{ij} = c_{ijkl}(\mathbf{y})X_{k,l}^{1\alpha\beta} - y_3 c_{ij\alpha\beta}(\mathbf{y}) \tag{4.1.13}$$

and verify that they are the solution of problem (4.1.11). For that, it is enough to verify that stresses (4.1.13) belong to Σ and satisfy (4.1.12).

Stresses (4.1.13) belong to Σ because $\mathbf{X}^{1\alpha\beta}$ is the solution of problem (4.1.1). Substituting (4.1.13) in (4.1.12) and integrating by parts, we obtain

$$\int_Y X_{i,j}^{1\alpha\beta}\eta_{ij}d\mathbf{y} = -\int_Y X_i^{1\alpha\beta}\eta_{ij,j}d\mathbf{y} + \int_S X_i^{1\alpha\beta}\eta_{ij}n_j d\mathbf{y} + \int_\omega X_i^{1\alpha\beta}\eta_{ij}n_j d\mathbf{y}. \tag{4.1.14}$$

Here, ω denotes the surfaces of domain Y perpendicular to the Oy_1y_2 plane (see Fig. 4.1).

The right-hand part of (4.1.14) is equal to zero by virtue of $\eta_{ij} \in \Sigma$; see the definition of set Σ (4.1.8). Then, stresses (4.1.13) satisfy (4.1.12).

To complete the proof, one must verify that $J_u(\mathbf{X}^{1\alpha\beta}) = J_\sigma(\sigma_{ij})$ with σ_{ij} given by (4.1.13) or (that is the same) that the following equality is valid:

$$\langle 2y_3 c_{ij\alpha\beta}(\mathbf{y})X_{i,j}^{1\alpha\beta} - c_{ijkl}(\mathbf{y})X_{i,j}^{1\alpha\beta}X_{k,l}^{1\alpha\beta}\rangle \tag{4.1.15}$$

$$= \langle c_{ijkl}^{-1}(\mathbf{y})[\sigma_{ij} + y_3 c_{ij\alpha\beta}(\mathbf{y})][\sigma_{kl} + y_3 c_{kl\alpha\beta}(\mathbf{y})]\rangle$$

with σ_{ij} given by (4.1.13).

Setting $\alpha\beta = \gamma\delta$, $\mu = \nu = 1$, in equality (4.1.4), we obtain

$$\langle y_3 c_{ij\alpha\beta}(\mathbf{y}) X_{i,j}^{1\alpha\beta} - c_{ijkl}(\mathbf{y}) X_{i,j}^{1\alpha\beta} X_{k,l}^{1\alpha\beta} \rangle = 0 . \tag{4.1.16}$$

Substituting stresses σ_{ij} given by (4.1.13) in (4.1.15), we obtain the equality,

$$\langle y_3 c_{ij\alpha\beta}(\mathbf{y}) X_{i,j}^{1\alpha\beta} \rangle = \langle c_{ijkl}(\mathbf{y}) X_{i,j}^{1\alpha\beta} X_{k,l}^{1\alpha\beta} \rangle ,$$

which is true by virtue of (4.1.16). Thus, equality (4.1.15) is true. Using Proposition 4.1, one can obtain the variational principles for bending stiffnesses. From (4.1.3), (4.1.5) when $\alpha\beta = \gamma\delta$, $\mu = \nu = 1$, we obtain the following relation between bending stiffnesses and the Lagrange functional:

$$A^2_{\alpha\beta\alpha\beta} = \langle y_3^2 c_{\alpha\beta\alpha\beta}(\mathbf{y}) \rangle - 2J_u(\mathbf{X}^{1\alpha\beta}) . \tag{4.1.17}$$

From Proposition 4.1 and (4.1.17), we obtain

$$A^2_{\alpha\beta\alpha\beta} = \langle y_3^2 c_{\alpha\beta\alpha\beta}(\mathbf{y}) \rangle - 2 \max_{\mathbf{u}\in V} J_u(\mathbf{u}) , \tag{4.1.18}$$

$$A^2_{\alpha\beta\alpha\beta} = \langle y_3^2 c_{\alpha\beta\alpha\beta}(\mathbf{y}) \rangle - 2 \min_{\sigma\in\Sigma} J_\sigma(\sigma) .$$

These are two variational principles (one in terms of displacements and the other in terms of stresses) for the bending stiffnesses of a nonhomogeneous plate.

When $\mathbf{u} \in V$ and $\sigma_{ij} \in \Sigma$ are arbitrary, we obtain from (4.1.18) the following two-sided estimates for bending stiffnesses:

$$\langle y_3^2 c_{\alpha\beta\alpha\beta}(\mathbf{y}) \rangle - 2J_u(\mathbf{u}) \geq A^2_{\alpha\beta\alpha\beta} \geq \langle y_3^2 c_{\alpha\beta\alpha\beta}(\mathbf{y}) \rangle - 2J_\sigma(\sigma) . \tag{4.1.19}$$

Taking into account (2.2), one can rewrite the right-hand side inequality from (4.1.19) in the following form:

$$A^2_{\alpha\beta\alpha\beta} \geq = \langle -c^{-1}_{ijkl}(\mathbf{y}) \sigma_{ij} \sigma_{kl} - 2\sigma_{\alpha\beta} y_3 \rangle . \tag{4.1.20}$$

Variational principles for in-plane stiffnesses

In this section, variational principles for in-plane stiffnesses $A^0_{\alpha\beta\alpha\beta}$ are derived. The in-plane stiffnesses are given by (4.1.3) with $\mu = \nu = 0$. To obtain the variational principles for in-plane stiffnesses, it is enough to repeat the computations presented in Sect. 2.2 with $\mu = \nu = 0$. Then, we obtain variational principles for in-plane stiffnesses $A^0_{\alpha\beta\alpha\beta}$. We present here only the final results.

The Lagrange functional is the following:

$$J_u(\mathbf{u}) = (1/2)\langle 2c_{ij\alpha\beta}(\mathbf{y})u_{i,j} - c_{ijkl}(\mathbf{y})u_{i,j}u_{k,l}\rangle . \qquad (4.1.21)$$

It is considered on the set V (4.1.6).

The Castigliano functional is the following:

$$J_\sigma(\sigma) = (1/2)\langle c^{-1}_{ijkl}(\mathbf{y})\sigma_{ij}\sigma_{kl} + 2\sigma_{\alpha\beta}y_3 + c_{\alpha\beta\alpha\beta}(\mathbf{y})\rangle . \qquad (4.1.22)$$

It is considered on the set Σ (4.1.8).

Proposition 4.2. *Under conditions C (Sect.* 1.2), *the following equations hold for the functionals determined by* (4.1.5) *and* (4.1.6):

$$\max_{\mathbf{u}\in V} J_u(\mathbf{u}) = \min_{\sigma\in\Sigma} J_\sigma(\sigma) = J_u(\mathbf{X}^{0\alpha\beta}), \qquad (4.1.23)$$

where $\mathbf{X}^{0\alpha\beta}$ *is the solution of problem* (4.1.1) *with* $\nu = 0$.

One can prove Proposition 4.2 similarly to the proof of Proposition 4. 1.

Using Proposition 4.2, one can obtain the following equalities:

$$A^0_{\alpha\beta\alpha\beta} = \langle c_{\alpha\beta\alpha\beta}(\mathbf{y})\rangle - 2\max_{\mathbf{u}\in V} J_u(\mathbf{u}), \qquad (4.1.24)$$

$$A^0_{\alpha\beta\alpha\beta} = \langle c_{\alpha\beta\alpha\beta}(\mathbf{y})\rangle - 2\min_{\sigma\in\Sigma} J_\sigma(\sigma).$$

These are two variational principles (one in terms of displacements and the other in terms of stresses) for in-plane stiffnesses of plate.

When $\mathbf{u} \in V$ and $\sigma_{ij} \in \Sigma$ are arbitrary, we obtain from (4.1.24) the following two-sided estimates for in-plane stiffnesses:

$$\langle c_{\alpha\beta\alpha\beta}(\mathbf{y})\rangle - 2J_u(\mathbf{u}) \geq A^0_{\alpha\beta\alpha\beta} \geq \langle c_{\alpha\beta\alpha\beta}(\mathbf{y})\rangle - 2J_\sigma(\sigma). \qquad (4.1.25)$$

The right-hand side inequality from (4.1.25) can be rewritten in the following form:

$$A^0_{\alpha\beta\alpha\beta} \geq = \langle -c^{-1}_{ijkl}(\mathbf{y})\sigma_{ij}\sigma_{kl} - 2\sigma_{\alpha\beta}\rangle . \qquad (4.1.26)$$

Variational principle for coupling stiffnesses

For a plate of arbitrary nonhomogeneous structure, there is no *a priori* way to introduce the "neutral" plane. A similar problem was discussed by Kolpakov (1998b) in connection with the "neutral" axis of a nonhomogeneous beam. As a result, one cannot decouple the in-plane strain from the bending, and the coupling stiffnesses must be considered.

The coupling stiffnesses $A^1_{\alpha\beta\alpha\beta}$ are given by (4.1.3) with $\mu+\nu=1$. They are not expressed in the form of a symmetrical quadratic functional. As a result, the technique used above cannot be applied directly to the coupling stiffnesses. In this section, we derive the variational principles for a combination of the in-plane, coupling, and bending stiffnesses.

We introduce the following Lagrange functional:

$$J_u(\mathbf{u}) = (1/2)\langle 2(y_3^\nu + h)c_{ij\alpha\beta}(\mathbf{y})u_{i,j} - c_{ijkl}(\mathbf{y})u_{i,j}u_{k,l}\rangle, \tag{4.1.27}$$

where h is an arbitrary nonzero real number.

The solution of the problem

$$J_u(\mathbf{u}) \to \max\,,\ \mathbf{u}\in V \tag{4.1.28}$$

is

$$\mathbf{X}^+ = \mathbf{X}^{1\alpha\beta} + h\mathbf{X}^{0\alpha\beta}, \tag{4.1.29}$$

where $\mathbf{X}^{\nu\alpha\beta}$ $(\nu=0,1)$ is the solution of (4.1.1).

Substituting (4.1.29) in (4.1.27), we obtain

$$J_u(\mathbf{X}^+) = (1/2)\langle 2y_3 c_{ij\alpha\beta}(\mathbf{y})X^{1\alpha\beta}_{i,j}(\mathbf{y}) - c_{ijkl}(\mathbf{y})X^{1\alpha\beta}_{i,j}(\mathbf{y})X^{1\alpha\beta}_{k,l}(\mathbf{y})\rangle$$

$$+ (1/2)\langle -2c_{ij\alpha\beta}(\mathbf{y})X^{0\alpha\beta}_{i,j}(\mathbf{y}) - c_{ijkl}(\mathbf{y})X^{0\alpha\beta}_{i,j}(\mathbf{y})X^{0\alpha\beta}_{k,l}(\mathbf{y})\rangle$$

$$+ (1/2)\langle -2c_{ij\alpha\beta}(\mathbf{y})X^{1\alpha\beta}_{i,j}(\mathbf{y}) + 2c_{ij\alpha\beta}(\mathbf{y})y_3\mathrm{X}^{0\alpha\beta}_{i,j}(\mathbf{y})$$

$$- c_{ijkl}(\mathbf{y})X^{1\alpha\beta}_{i,j}(\mathbf{y})X^{0\alpha\beta}_{k,l}(\mathbf{y}) - c_{ijkl}(\mathbf{y})\mathrm{X}^{0\alpha\beta}_{i,j}(\mathbf{y})\mathrm{X}^{1\alpha\beta}_{k,l}(\mathbf{y})\rangle\,.$$

In accordance with (4.1.17), the first term on the right-hand side is equal to $(1/2)(-2A^2_{\alpha\beta\alpha\beta} + \langle c_{\alpha\beta\alpha\beta}(\mathbf{y})y_3^2\rangle)$. The second term is equal to $(1/2)[-A^0_{\alpha\beta\alpha\beta} + \langle c_{\alpha\beta\alpha\beta}(\mathbf{y})\rangle h^2]$. The third term can be written as

$$(1/2)\langle -c_{ij\alpha\beta}(\mathbf{y})X^{1\alpha\beta}_{i,j}(\mathbf{y})+c_{ij\alpha\beta}(\mathbf{y})y_3\mathrm{X}^{0\alpha\beta}_{i,j}(\mathbf{y})\rangle h$$

$$+(1/2)\langle[-c_{ij\alpha\beta}(\mathbf{y})-c_{ijkl}(\mathbf{y})\mathrm{X}^{0\alpha\beta}_{k,l}(\mathbf{y})]\mathrm{X}^{1\alpha\beta}_{i,j}(\mathbf{y})\rangle h$$

$$+(1/2)\langle[y_3 c_{ij\alpha\beta}(\mathbf{y})-c_{ijkl}(\mathbf{y})\mathrm{X}^{1\alpha\beta}_{k,l}(\mathbf{y})]\mathrm{X}^{0\alpha\beta}_{i,j}(\mathbf{y})\rangle h\,.$$

Let us consider this expression. In accordance with (4.1.3), the first term is equal to $-(A^1_{\alpha\beta\alpha\beta}+\langle c_{\alpha\beta\alpha\beta}(\mathbf{y})y_3\rangle)h$. By virtue of (4.1.4), the two last terms are equal to zero.

As a result, we obtain the equality

$$A^2_{\alpha\beta\alpha\beta}+2hA^1_{\alpha\beta\alpha\beta}+h^2A^0_{\alpha\beta\alpha\beta} \tag{4.1.30}$$

$$=\langle c_{\alpha\beta\gamma\delta}(\mathbf{y})y_3^2\rangle+2\langle c_{\alpha\beta\gamma\delta}(\mathbf{y})y_3\rangle h+\langle c_{\alpha\beta\gamma\delta}(\mathbf{y})\rangle h^2-2J_u(\mathbf{X}^+)\,.$$

From (4.1.30) and (4.1.28), we obtain

$$A^2_{\alpha\beta\alpha\beta}+2hA^1_{\alpha\beta\alpha\beta}+h^2A^0_{\alpha\beta\alpha\beta}=\langle c_{\alpha\beta\alpha\beta}(\mathbf{y})\rangle h^2-2\max_{\mathbf{u}\in V}J_u(\mathbf{u})\,. \tag{4.1.31}$$

We can derive the following dual (Castigliano-type) variational principle:

$$A^2_{\alpha\beta\alpha\beta}+2hA^1_{\alpha\beta\alpha\beta}+h^2A^0_{\alpha\beta\alpha\beta}=\langle c_{\alpha\beta\alpha\beta}(\mathbf{y})\rangle h^2-2\min_{\sigma\in\Sigma}J_\sigma(\sigma)\,, \tag{4.1.32}$$

where

$$J_\sigma(\sigma)=(1/2)\langle c^{-1}_{ijkl}(\mathbf{y})\sigma_{ij}\sigma_{kl}+2\sigma_{\alpha\beta}(y_3+h)+(y_3+h)^2c_{\alpha\beta\alpha\beta}(\mathbf{y})\rangle\,. \tag{4.1.33}$$

Variational principles (4.1.31) and (4.1.32) are a pair of variational principles for the expression $A^2_{\alpha\beta\alpha\beta}+2hA^1_{\alpha\beta\alpha\beta}+h^2A^0_{\alpha\beta\alpha\beta}$. Using variational principles (4.1.31) and (4.1.32), one can compute the coupling stiffnesses if the in-plane and bending stiffnesses are computed.

From (4.1.31) and (4.1.32), one can derive two-sided estimate,

$$\langle c_{\alpha\beta\alpha\beta}(\mathbf{y})\rangle h^2-2J_u(\mathbf{u}) \tag{4.1.34}$$

$$\geq A^2_{\alpha\beta\alpha\beta}+2hA^1_{\alpha\beta\alpha\beta}+h^2A^0_{\alpha\beta\alpha\beta}\geq\langle c_{\alpha\beta\alpha\beta}(\mathbf{y})\rangle h^2-2J_\sigma(\sigma)\,.$$

Variational principles for combinations of stiffnesses

Variational principles for stiffnesses $A^{\mu+\nu}_{\alpha\beta\gamma\delta}$ with arbitrary indexes

The technique used above provides variational principles for stiffnesses $A^{\mu+\nu}_{\alpha\beta\gamma\delta}$ with indexes of the form $\alpha\beta\alpha\beta$. In this section, we derive variational principles for combinations of stiffnesses.

We consider the following Lagrange functional:

$$J_u(\mathbf{u}) = (1/2)\langle 2(-1)^\nu C_{\alpha\beta} y_3^\nu c_{ij\alpha\beta}(\mathbf{y})u_{i,j} - c_{ijkl}(\mathbf{y})u_{i,j}u_{k,l}\rangle \tag{4.1.35}$$

($\nu = 0,1$), where $C_{\alpha\beta}$ are arbitrary constants.

The solution of the problem

$$J_u(\mathbf{u}) \to \max, \ \mathbf{u} \in V$$

is

$$\mathbf{X}^+ = C_{\alpha\beta}\mathbf{X}^{\nu\alpha\beta}, \tag{4.1.36}$$

where $\mathbf{X}^{\nu\alpha\beta}$ ($\nu = 0,1$) means the solution of problem (4.1.1).

Substituting (4.1.36) in (4.1.35), and after algebraic transformations, we obtain the following equality:

$$J_u(\mathbf{X}^+) = C_{\alpha\beta}C_{\gamma\delta}(\langle c_{\alpha\beta\gamma\delta}(\mathbf{y})y_3^{2\nu}\rangle - A^{2\nu}_{\alpha\beta\gamma\delta}) \quad (\nu = 0,1). \tag{4.1.37}$$

Comparing (4.1.37) and (4.1.35), (4.1.36), we obtain

$$C_{\alpha\beta}C_{\gamma\delta}A^{2\nu}_{\alpha\beta\gamma\delta} = C_{\alpha\beta}C_{\gamma\delta}\langle c_{\alpha\beta\gamma\delta}(\mathbf{y})y_3^{2\nu}\rangle - 2\max_{\mathbf{u}\in V} J_u(\mathbf{u}), \tag{4.1.38}$$

where $J_u(\mathbf{u})$ is defined by (4.1.35).

The Castigliano functional has the form,

$$J_\sigma(\sigma) = (1/2)\langle c^{-1}_{ijkl}(\mathbf{y})\sigma_{ij}\sigma_{kl} + 2C_{\alpha\beta}\sigma_{\alpha\beta}y_3 + C_{\alpha\beta}C_{\gamma\delta}c_{\alpha\beta\gamma\delta}(\mathbf{y})\rangle .$$

Proposition 4.4. *Under conditions C, the following equation holds for the functionals determined by* (4.1.35) *and* (4.1.38):

$$\max_{\mathbf{u}\in V} J_u(\mathbf{u}) = \min_{\sigma\in\Sigma} J_\sigma(\sigma).$$

From (4.1.38) and Proposition 4.4,

$$C_{\alpha\beta}C_{\gamma\delta}A^{2\nu}_{\alpha\beta\gamma\delta} = C_{\alpha\beta}C_{\gamma\delta}\langle c_{\alpha\beta\gamma\delta}(\mathbf{y})y_3^{2\nu}\rangle - 2\min_{\sigma\in\Sigma} J_\sigma(\sigma). \tag{4.1.39}$$

Variational principles (4.1.38) and (4.1.39) are a pair of variational principles for the expression $C_{\alpha\beta}C_{\gamma\delta}A^{2\nu}_{\alpha\beta\gamma\delta}$.

An example

We write variational principle for the stiffnesses $A^{2\nu}_{1122}$. Set $C_{11} = C_{22} = 1$, and $C_{\alpha\beta} = 0$ otherwise. We obtain from (4.1.38) the following:

$$A^{2\nu}_{1212} = \langle c_{1212}(\mathbf{y})y_3^{2\nu}\rangle - 2\min_{\sigma\in\Sigma} J_\sigma(\sigma) \quad (\nu = 0,1),$$

where the Lagrange functional has the form,

$$J_u(\mathbf{u}) = (1/2)\langle 2(-1)^\nu y_3^\nu [c_{ij11}(\mathbf{y}) + c_{ij22}(\mathbf{y})]u_{i,j} - c_{ijkl}(\mathbf{y})u_{i,j}u_{k,l}\rangle.$$

Variational principles for coupling stiffnesses

The principles obtained above can be applied only to in-plane ($\nu = 0$) or bending ($\nu = 1$) stiffnesses. We consider the coupling stiffnesses.

We consider the following Lagrange functional:

$$J_u(\mathbf{u}) = (1/2)\langle 2C_{\alpha\beta}(y_3 - h)c_{ij\alpha\beta}(\mathbf{y})u_{i,j} - c_{ijkl}(\mathbf{y})u_{i,j}u_{k,l}\rangle \quad (\nu = 0,1), \tag{4.1.40}$$

where $C_{\alpha\beta}$ and h are arbitrary real numbers.

The following equality holds:

$$C_{\alpha\beta}C_{\gamma\delta}(A^2_{\alpha\beta\gamma\delta} + 2hA^1_{\alpha\beta\gamma\delta} + h^2A^0_{\alpha\beta\gamma\delta}) \tag{4.1.41}$$

$$= C_{\alpha\beta}C_{\gamma\delta}[\langle c_{\alpha\beta\gamma\delta}(\mathbf{y})y_3^2\rangle - 2\langle c_{\alpha\beta\gamma\delta}(\mathbf{y})y_3\rangle h + \langle c_{\alpha\beta\gamma\delta}(\mathbf{y})\rangle h^2] - 2\max_{\mathbf{u}\in V} J_u(\mathbf{u}).$$

It can be derived similarly to that used to obtain equalities (4.1.38) and (4.1.31).

The Castigliano functional has the form,

$$J_\sigma(\sigma) = \frac{1}{2}\langle c^{-1}_{ijkl}(\mathbf{y})\sigma_{ij}\sigma_{kl} + 2C_{\alpha\beta}\sigma_{\alpha\beta}(y_3 - h) + C_{\alpha\beta}C_{\gamma\delta}(y_3 - h)^2 c_{\alpha\beta\gamma\delta}(\mathbf{y})\rangle. \tag{4.1.42}$$

Proposition 4.5. *Under conditions C, the following equation holds for functionals* (4.1.40) *and* (4.1.42):

$$\max_{\mathbf{u}\in V} J_u(\mathbf{u}) = \min_{\sigma\in\Sigma} J_\sigma(\sigma). \tag{4.1.43}$$

Propositions 4.4 and 4.5 can be proved similarly to that used to prove Proposition 4.1.

From (4.1.41) and Proposition 4.5,

$$C_{\alpha\beta}C_{\gamma\delta}(A^2_{\alpha\beta\gamma\delta} + 2hA^1_{\alpha\beta\gamma\delta} + h^2 A^0_{\alpha\beta\gamma\delta}) \tag{4.1.44}$$

$$= C_{\alpha\beta}C_{\gamma\delta}[\langle c_{\alpha\beta\gamma\delta}(\mathbf{y})y_3^2\rangle - 2\langle c_{\alpha\beta\gamma\delta}(\mathbf{y})y_3\rangle h + \langle c_{\alpha\beta\gamma\delta}(\mathbf{y})\rangle h^2] - 2\min_{\sigma\in V} J_\sigma(\sigma).$$

The variational principles (4.1.41) and (4.1.43) involve the coupling stiffnesses and are a pair of variational principles for the expression,

$$C_{\alpha\beta}C_{\gamma\delta}(A^2_{\alpha\beta\gamma\delta} + 2hA^1_{\alpha\beta\gamma\delta} + h^2 A^0_{\alpha\beta\gamma\delta}).$$

Using the combinations introduced above, we can write the variational principles for the stiffness with any given index. We consider, the coefficients $C_{\alpha\beta}$ given by the formula,

$$C_{\alpha\beta} = 1 \text{ if } \alpha\beta = PQ \text{ and } \gamma\delta = KL\text{, and } C_{\alpha\beta} = 0 \text{ otherwise.}$$

Here $PQKL$ is a given index. With regard to the symmetry of the stiffnesses, $C_{\alpha\beta}C_{\gamma\delta}A^{2\nu}_{\alpha\beta\gamma\delta} = 2A^{2\nu}_{PQKL}$. For the same coefficients $C_{\alpha\beta}$, the expression $C_{\alpha\beta}C_{\gamma\delta}(A^2_{\alpha\beta\gamma\delta} + 2hA^1_{\alpha\beta\gamma\delta} + h^2 A^0_{\alpha\beta\gamma\delta})$ becomes $2(A^2_{PQKL} + 2hA^1_{PQKL} + h^2 A^0_{PQKL})$. It is easily to write the corresponding Lagrange and Castigliano functionals.

Some estimates for the stiffnesses of a nonhomogeneous plate

We present here some estimates following from the variational principles obtained above.

The Voight-like upper bound for stiffnesses

Putting $\mathbf{u} = 0$ into (4.1.19), (4.1.25), and (4.1.34) (it is clear that $0 \in V$), we obtain the following estimates for stiffnesses:

$$A^2_{\alpha\beta\alpha\beta} \le \langle c_{\alpha\beta\alpha\beta}(\mathbf{y})y_3^2\rangle, \tag{4.1.45}$$

$$A^{0}_{\alpha\beta\alpha\beta} \leq \langle c_{\alpha\beta\alpha\beta}(\mathbf{y})\rangle ,$$

$$A^{2}_{\alpha\beta\alpha\beta} + 2hA^{1}_{\alpha\beta\alpha\beta} + h^{2}A^{0}_{\alpha\beta\alpha\beta} \leq \langle c_{\alpha\beta\alpha\beta}(\mathbf{y})\rangle h^{2} .$$

We consider a plate made of isotropic materials. Expressing $c_{\alpha\beta\alpha\beta}$ through Young's modulus $E(\mathbf{y})$ and Poisson's ratio $\nu(\mathbf{y})$, one can write (4.1.45) in the form,

$$\mathrm{A}^{2}_{\alpha\alpha\alpha\alpha} \leq \left\langle \frac{y_3^2(1-\nu(\mathbf{y}))E(\mathbf{y})}{(1-2\nu(\mathbf{y}))(1+\nu(\mathbf{y}))} \right\rangle \ (\alpha = 1, 2), \qquad (4.1.46)$$

$$\mathrm{A}^{2}_{1212} \leq \left\langle \frac{y_3^2\nu(\mathbf{y})E(\mathbf{y})}{(1-2\nu(\mathbf{y}))(1+\nu(\mathbf{y}))} \right\rangle ,$$

$$\mathrm{A}^{0}_{\alpha\alpha\alpha\alpha} \leq \left\langle \frac{(1-\nu(y))E(y)}{(1-2\nu(y))(1+\nu(y))} \right\rangle ,$$

$$\mathrm{A}^{0}_{1212} \leq \left\langle \frac{\nu(\mathbf{y})E(\mathbf{y})}{(1-2\nu(\mathbf{y}))(1+\nu(\mathbf{y}))} \right\rangle ,$$

$$A^{2}_{\alpha\alpha\alpha\alpha} + 2hA^{1}_{\alpha\alpha\alpha\alpha} + h^{2}A^{0}_{\alpha\alpha\alpha\alpha} \leq \left\langle \frac{(1-\nu(\mathbf{y}))E(\mathbf{y})}{(1-2\nu(\mathbf{y}))(1+\nu(\mathbf{y}))} \right\rangle h^{2} ,$$

$$A^{2}_{1212} + 2hA^{1}_{1212} + h^{2}A^{0}_{1212} \leq \left\langle \frac{\nu(\mathbf{y})E(\mathbf{y})}{(1-2\nu(\mathbf{y}))(1+\nu(\mathbf{y}))} \right\rangle h^{2} .$$

The estimates (4.1.46) are plate analogies of Voight's bound (Voight, 1928).

Refined upper bound

We consider the set $C^{1}([A,B])$ of differentiable functions $\mathbf{f}(y_3)$ of unique variable y_3. Here, $A = \min_{\mathbf{y}\in Y} y_3$, $B = \max_{\mathbf{y}\in Y} y_3$. It is clear that $C^{1}([A,B]) \subset V$.

Putting $\mathbf{u} = \mathbf{f}(y_3)$ into (4.1.19), (4.1.25), we obtain the following inequality:

$$A^{2\nu}_{\alpha\beta\alpha\beta} \le \langle c_{\alpha\beta\alpha\beta}(\mathbf{y}) y_3^{2\nu} \rangle - 2 \max_{\mathbf{f}\in C^1} J_u(\mathbf{f}) \ (\nu=0,\,1),$$

where $C^1 = C^1([A,B])$.

For simplicity, we consider a plate made of isotropic materials. In this case, one can take $f_1 = f_2 = 0$, $f_3 = f_3(y_3)$, and the Lagrange functional (4.1.5) takes the form, $J_u(\mathbf{f}) = (1/2) \le \langle 2y_3^{\nu} c_{33\alpha\beta}(\mathbf{y}) f' - c_{33\alpha\beta}(\mathbf{y}) f'^2 \rangle$, where the prime means the derivation with respect to y_3.

Euler's equation for the functional $J_u(\mathbf{f})$ has the form,

$$\int_Y [c_{33\alpha\beta}(\mathbf{y}) y_3^{\nu} - c_{3333} f'] \eta(y_3) \mathrm{d}\mathbf{y} = 0 \quad \text{for any } \eta(y_3) \in C^1. \tag{4.1.47}$$

We introduce the functions

$$g_{mn}(y_3) = \int_{S(y_3)} \mathrm{c}_{33\mathrm{mn}}(y_1, y_2, y_3) \mathrm{d}y_1 \mathrm{d}y_2,$$

where $S(y_3) = Y \cap \{y_3 = const\}$ means the intersection of periodicity cell Y and the plane $\{y_3 = const\}$. One can derive the following equation from (4.1.47):

$$g_{\alpha\beta}(y_3) y_3^{\nu} - g_{33}(y_3) f' = 0.$$

The solution of this equation is

$$f'(y_3) = \frac{g_{\alpha\beta}(y_3) y_3^{\nu}}{g_{33}(y_3)}. \tag{4.1.48}$$

Substituting (4.1.48) in the formula for calculating $J_u(\mathbf{f})$, we obtain

$$\max_{f\in C^1} J_u(\mathbf{f}) = \left\langle \frac{g_{\alpha\beta}(y_3) y_3^{2\nu}}{g_{33}(y_3)} \right\rangle. \tag{4.1.49}$$

One can derive the following estimate from (4.1.49):

$$A^{2\nu}_{\alpha\beta\alpha\beta} \le \left\langle \frac{c_{\alpha\beta\alpha\beta}(\mathbf{y}) - [g_{\alpha\beta}(y_3)]^2}{g_{33}(y_3) y_3^{2\nu}} \right\rangle \ (\nu = 0, 1). \tag{4.1.50}$$

For a laminated plate, (4.1.48) gives the solution of cellular problem (4.1.1) (see, e.g., Kalamkarov and Kolpakov, 1997). Thus, for laminated plates, the right-hand side of (4.1.50) is equal to stiffness $A^{2\nu}_{\alpha\beta\alpha\beta}$. This means that (5.1.50) is a sharp estimate.

An example

Let the functions $g_{mn}(y_3)$ not depend on y3. In this case, $g_{mn}(y_3) = \langle c_{33mn}(\mathbf{y})\rangle$. Then the right-hand side of equality (4.1.49) takes the form,

$$(1/2)\frac{\langle c_{33\alpha\beta}(\mathbf{y})\rangle^2 \langle y_3^{2\nu}\rangle}{\langle c_{3333}(\mathbf{y})\rangle} ,$$

and estimate (4.1.50) takes the form,

$$A^{2\nu}_{\alpha\beta\alpha\beta} \le \left(\langle c_{\alpha\beta\alpha\beta}(\mathbf{y})\rangle - \frac{\langle c_{22\alpha\beta}(\mathbf{y})\rangle^2}{\langle c_{3333}(\mathbf{y})\rangle} \right) \langle y_3^{2\nu}\rangle . \tag{4.1.51}$$

If the materials from which the plate is made are isotropic and Poisson's ratios ν of all the components are the same, one can write (4.1.51) in the terms of Young's modulus E and the Poisson's ratios ν as follows:

$$A^0_{\alpha\beta\alpha\beta} \le \frac{\langle E(\mathbf{y})\rangle}{1-\nu^2}, \quad A^2_{\alpha\beta\alpha\beta} \le \frac{\langle E(\mathbf{y})\rangle\langle y_3^2\rangle}{1-\nu^2} .$$

In accordance with the definition of function $g_{mn}(y_3)$, the condition "the functions $g_{mn}(y_3)$ do not depend on y_3" is equivalent to the condition "the volume ratios of components in the cross section $S(y_3)$ do not depend on y_3." There are many composite structures satisfying this condition.

Estimate for a plate having planar lateral surfaces. Lower bound

Let us consider a non-homogeneous plate of symmetrical structure having planar lateral surfaces. Consider the stress tensor of the form,

$$\sigma_{\alpha\beta} = C_{\alpha\beta} f(y_3), \quad \sigma_{i3} = 0 , \tag{4.1.52}$$

where $C_{\alpha\beta}$ are arbitrary real numbers, α, β $(\alpha, \beta = 1, 2)$ are fixed indexes, $f(y_3)$ is an arbitrary function of variable y_3.

For the stresses given by (4.1.52), $\sigma_{ij,j} = \delta_{i\alpha}\delta_{j\beta}C_{\alpha\beta}f(y_3)_{,j}$ in Y. Taking into account that in the case under consideration, the normal vector has the form $\mathbf{n} = (0,0,1)$, $\sigma_{ij}n_j = \sigma_{i3} = 0$ on S. Then, $\sigma_{ij} \in \Sigma$, and one can use these stresses in estimates (4.1.26) and (4.1.20). For stresses (4.1.52), the right-hand sides of (4.1.20) and (4.1.26) take the form,

$$-C_{\chi\delta}C_{\mu\nu}\langle c^{-1}_{\chi\delta\mu\nu}(\mathbf{y})f(y_3)^2\rangle + 2C_{\alpha\beta}\langle f(y_3)y_3^n\rangle, \qquad (4.1.53)$$

$n = 0$ for (4.1.26), and $n = 1$ for (4.1.20).

There are no restrictions on $C_{\chi\delta}$. Then, one can maximize (4.1.53) with respect to $C_{\chi\delta}$. Euler's equation for (4.1.53) is

$$-C_{\mu\nu}\langle c^{-1}_{\chi\delta\mu\nu}(\mathbf{y})f(y_3)^2\rangle + \delta_{\chi\alpha}\delta_{\delta\beta}\langle f(y_3)y_3^n\rangle = 0. \qquad (4.1.54)$$

The solution of (4.1.54) is

$$C_{\mu\nu} = \langle c^{-1}_{\mu\nu\alpha\beta}(\mathbf{y})f(y_3)^2\rangle^{-1}\langle f(y_3)y_3^n\rangle,$$

where $\langle c^{-1}_{\mu\nu\alpha\beta}(\mathbf{y})f(y_3)^2\rangle^{-1}$ means the tensor inverse with respect to $\langle c^{-1}_{\mu\nu\alpha\beta}(\mathbf{y})f(y_3)^2\rangle$. Substituting this solution in (4.1.53), we obtain the following estimates:

$$A^{2\nu}_{\alpha\beta\alpha\beta} \geq \langle c^{-1}_{\alpha\beta\alpha\beta}(\mathbf{y})f(y_3)^2\rangle^{-1}\langle f(y_3)y_3^n\rangle. \qquad (4.1.55)$$

Reuss-like bound

Consider estimate (4.1.55) with $f(y_3) = 1$. It is the case of uniform stresses, the case analyzed for 3-D composites by Reuss (1929). For $\nu = 0$, we obtain the plate analog of Reuss's bound for in-plane stiffnesses,

$$A^{0}_{\alpha\beta\alpha\beta} \geq \langle c^{-1}_{\alpha\beta\alpha\beta}(\mathbf{y})\rangle^{-1}. \qquad (4.1.56)$$

For $n = 1$ and $f(y_3) = 1$, Reuss's bound for bending stiffnesses of a symmetrical plate is trivial: $A^{0}_{\alpha\beta\alpha\beta} \geq 0$ because of $\langle y_3\rangle = 0$ due to plate symmetry. The first nontrivial case holds if $n = 1$ and $f(y_3) = y_3$. In this case, we obtain the following estimate for bending stiffnesses:

$$A^{0}_{\alpha\beta\alpha\beta} \geq \langle c^{-1}_{\alpha\beta\alpha\beta}(\mathbf{y})y_3^2\rangle^{-1}\langle y_3^2\rangle^2. \qquad (4.1.57)$$

Estimate (4.1.57) may be called a "Reuss-type" estimate for the bending stiffnesses of the plate.

We consider a plate made of isotropic materials. In this case (4.1.56), (4.1.57) give

$$A^2_{\alpha\alpha\alpha\alpha} \geq \frac{\langle y_3^2\rangle^2}{\left\langle \dfrac{y_3^2}{E(\mathbf{y})} \right\rangle}, \tag{4.1.58}$$

$$A^2_{1212} \geq \frac{\langle y_3^2\rangle^2}{\left\langle \dfrac{y_3^2[1+\nu(\mathbf{y})]}{E(\mathbf{y})} \right\rangle},$$

$$A^0_{\alpha\alpha\alpha\alpha} \geq \frac{1}{\left\langle \dfrac{1}{E(\mathbf{y})} \right\rangle},$$

$$A^2_{1212} \geq \frac{1}{\left\langle \dfrac{[1+\nu(\mathbf{y})]}{E(\mathbf{y})} \right\rangle} \quad (\alpha = 1,\ 2)\ .$$

The difference between Reuss's estimates for in-plane stiffnesses and "Reuss-type" estimates for bending stiffnesses is clearly seen.

Refined estimates

Let us analyze (4.1.55) in the general case. The function $f(y_3)$ is an arbitrary function and one can maximize the right-hand side of (4.1.55) with respect to $f(y_3)$. Euler's equation for the functional on the right-hand side of (4.1.55) is as follows:

$$-M^{-1}\{g_{\alpha\beta}(y_3)g_{\alpha\beta}(y_3)\}M^{-1}\langle f(y_3)y_3^n\rangle + M\langle g_{\alpha\beta}(y_3)y_3^n\rangle y_3 = 0\,, \tag{4.1.59}$$

where $g_{\alpha\beta}(y_3) = \int\limits_{S(y_3)} c^{-1}_{\alpha\beta\alpha\beta}(y_1,y_2,y_3)dy_1dy_2$ and $M = \langle c^{-1}_{\alpha\beta\alpha\beta}(\mathbf{y})f(y_3)\rangle$. In the case under consideration, $c^{-1}_{\alpha\beta\alpha\beta}$ can depend on all variables y_1, y_2, y_3; $S(y_3) = Y \cap \{y_3 = const\}$.

The solution of (4.1.59) has the form, $f(y_3) = \text{const}\dfrac{y_3^n}{f(y_3)}$. Substituting this expression in the right-hand side of (4.1.55), we obtain the following estimate:

$$A^{2\nu}_{\alpha\beta\alpha\beta} \geq \left\langle \frac{y_3^{2\nu}}{g_{\alpha\beta}(y_3)} \right\rangle \quad (\nu = 0,\ 1).$$

Unidirectional plates

Let periodicity cell Y be a cylinder whose axis is parallel to the Oy_1 axis (Fig.4.2).

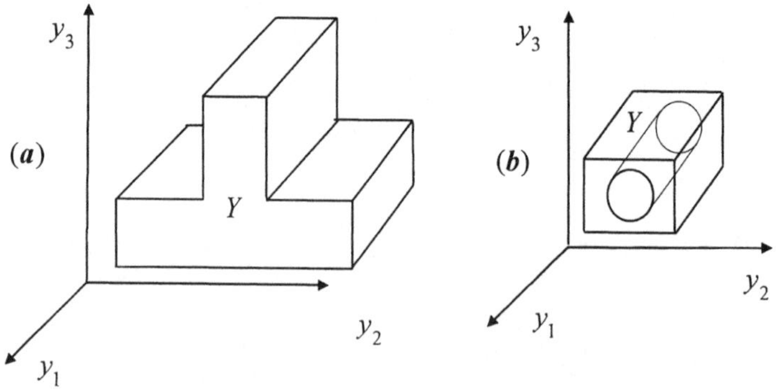

Fig. 4.2. Cylindrical periodicity cells: (***a***) ribbed cell and (***b***) fiber-reinforced cell

In the case under consideration, $n_1 = 0$ on S, and the stress tensor of the form, $\sigma_{11} = \sigma_{11}(y_2, y_3) \neq 0$, $\sigma_{ij} = 0$ if $ij \neq 11$, belongs to Σ. Let us consider stiffness A^2_{1111}. The right-hand sides of inequalities (4.1.20) and (4.1.26) take the form,

$$\langle -\frac{\sigma_{11}^2}{E(y_2, y_3)} - 2\sigma_{11} y_3^{\nu} \rangle, \tag{4.1.60}$$

$\nu = 0$ for (4.1.20) and $\nu = 1$ for (4.1.26).

Euler's equation for (4.1.60) has the form, $-\sigma_{11} / E(y_2, y_3) - y_3^{\nu} = 0$. Its solution is $\sigma_{11} = -E(y_2, y_3) y_3^{\nu}$. Substituting this solution in (4.1.60), we obtain

$$A^2_{1111} \geq \langle E(y_2, y_3) y_3^2 \rangle, \tag{4.1.61}$$

$$A^0_{1111} \geq \langle E(y_2, y_3) \rangle .$$

It is of interest that the Voight-type estimates are lower bounds here.

An example

We derived above the following estimates for the stiffnesses of a plate having planar lateral surfaces:

$$\frac{\langle y_3^2 \rangle^2}{\langle y_3^2 E(\mathbf{y}) \rangle} \leq A^2_{1111} \leq \left\langle \frac{y_3^2 [1-\nu(\mathbf{y})] E(\mathbf{y})}{[1-2\nu(\mathbf{y})][1+\nu(\mathbf{y})]} \right\rangle .$$

These estimates can be used for laminated plates. We consider a laminated plate (symmetrical with respect to its middle plane) formed from 400 layers and compute the exact value of bending stiffness, using the formula $A^2_{1111} = \dfrac{\langle y_3^2 E(\mathbf{y}) \rangle}{1-\nu^2}$ obtained from Kalamkarov and Kolpakov (1997), and estimates. In the computations, Poisson's ratio of layers was varied from 0 to 0.45. The Young's moduli of layers were randomly taken from the set $\{1,2,3,4,5\}$. The results of the computations are presented in Table 1.

Table 1. The bending stiffness of a laminated plate. Legend: Low, Upp - lower and upper estimates, Ex - exact value

ν	0.00	0.05	0.10	0.15	0.20	0.25	0.30	0.35	0.40	0.45
Low	2.06	2.06	1.90	1.90	1.90	1.90	1.75	1.75	1.75	1.75
Ex	2.64	2.65	2.48	2.51	2.56	2.62	2.65	2.70	2.73	2.88
Upp	2.64	2.65	2.51	2.59	2.73	2.94	3.10	3.69	4.93	8.72

4.2 Variational Principles for Stiffnesses of Nonhomogeneous Beams

In this section, variational principles and two-sided estimates for a nonhomogeneous beam of periodic structure are derived on the basis of the asymptotic method and analysis of Lagrange and Castigliano functionals for cellular problems.

Formulation of the problem

We consider an elastic body of small diameter formed by periodic repetition of periodicity cell εY along the Ox_1 axis described in Sect. 3.2. The tensor of local elastic constants of the body considered has the form, $c_{ijmn}(\mathbf{x}/\varepsilon)$, where $c_{ijmn}(\mathbf{y})$ are periodic functions with respect to y_1 with period T. Here $\mathbf{y} = \mathbf{x}/\varepsilon$ is the local variable, and T is the projection of Y on the Oy_1 axis (Fig. 4.3). It is assumed that the local elastic constants satisfy conditions C from Sect. 1.2.

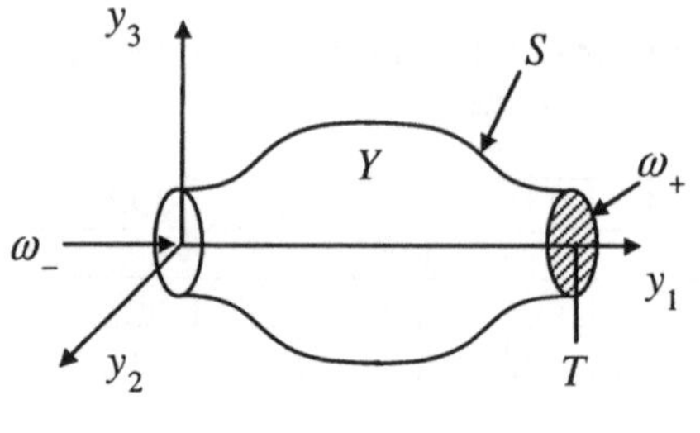

Fig. 4.3. Periodicity cell of beam

Variational principles and estimates for bending stiffnesses

Beam stiffnesses can be calculated as follows. At the first step, the so-called cellular problem is solved. In the case under consideration, the cellular problem has the form,

$$\begin{cases} [c_{ijmn}(\mathbf{y})X^{\alpha}_{m,n} + c_{ij11}(\mathbf{y})y_{\alpha}]_{,j} = 0 \quad \text{in } Y, \\ [c_{ijmn}(\mathbf{y})X^{\alpha}_{m,n} + c_{ij11}(\mathbf{y})y_{\alpha}]n_j = 0 \quad \text{on } S, \\ \mathbf{X}^{\alpha}(\mathbf{y}) \text{ is periodic in } y_1 \text{ with periodicity cell } T. \end{cases} \tag{4.2.1}$$

Here, S denotes the lateral boundary of the periodicity cell (it can include the surfaces of pores, if they exist); $\mathbf{n}$ is the normal vector to S.

All problems in this section are written in "fast" variables $\mathbf{y}$. The subscript $,j$ means $\partial / \partial y_j$. The Latin indexes take values 1, 2, 3; the Greek indexes take values 2, 3.

After solving problem (4.2.1), the stiffnesses $A^2_{\alpha\alpha}$ are calculated in accordance with the formula,

$$A^2_{\alpha\alpha} = \langle [c_{1111}(\mathbf{y})y_{\alpha} + c_{11mn}(\mathbf{y})X^{\alpha}_{m,n}(\mathbf{y})]y_{\alpha} \rangle . \tag{4.2.2}$$

In the paper by Kolpakov (1994) the following formula for a uniform (cylindrical) beam was presented:

$$A^2_{\alpha\alpha} = \langle c_{ijkl}(\mathbf{y})[X^{\alpha}_{k,l} - y_{\alpha}\delta_{k1}\delta_{l1}][X^{\alpha}_{i,j} - y_{\alpha}\delta_{i1}\delta_{j1}] \rangle , \tag{4.2.3}$$

giving stiffnesses in the form of a quadratic functional.

We prove that (4.2.3) remains valid for nonuniform beams. For that, let us derive (4.2.3) directly from (4.2.1), (4.2.2). Multiplying the first equation in (4.2.1) by X^{α}_i and integrating by parts with regard to the boundary and periodicity conditions from (4.2.1), we obtain the following equality:

$$\langle c_{ijkl}(\mathbf{y})[X^{\alpha}_{k,l} - y_{\alpha}\delta_{k1}\delta_{l1}]X^{\alpha}_{i,j} \rangle = 0 . \tag{4.2.4}$$

Adding (4.2.4) and (4.2.2), we obtain (4.2.3).

Torsional stiffness

Torsional stiffness B is computed similarly (see Sect. 3.2). The corresponding cellular problem is the following:

$$\begin{cases} [c_{ijmn}(\mathbf{y})X^3_{m,n} + c_{ij\gamma 1}(\mathbf{y})s_\gamma y_\Gamma]_{,j} = 0 & \text{in } Y, \\ [c_{ijmn}(\mathbf{y})X^3_{m,n} + c_{ij\gamma 1}(\mathbf{y})s_\gamma y_\Gamma]n_j = 0 & \text{on } S, \\ \mathbf{X}^3(\mathbf{y}) \text{ is periodic in } y_1 \text{ with period } T. \end{cases} \tag{4.2.5}$$

As above, $\Gamma = 3$ if $\gamma = 2$, and $\Gamma = 2$ if $\gamma = 3$; $s_2 = 1$; $s_3 = -1$.

After solving problem (4.2.5), torsional stiffness B is calculated in accordance with the formula,

$$B = \langle y_\Gamma s_\gamma [c_{\gamma 1kl}(\mathbf{y})X_{k,l} + y_\Gamma s_\gamma c_{\gamma 1\gamma 1}(\mathbf{y})]\rangle . \tag{4.2.6}$$

Formula (4.2.6) can be written in the form of a quadratic functional:

$$B = \langle c_{ijkl}(\mathbf{y})[X_{k,l} + y_\Gamma s_\gamma \delta_{k\gamma}\delta_{l1}][X_{i,j} + y_\Gamma s_\gamma \delta_{i\gamma}\delta_{j1}]\rangle . \tag{4.2.7}$$

Formula (4.2.7) can be derived from (4.2.5), (4.2.6), as above.

Variational principles and estimates for bending stiffnesses

Let us derive the variational principle, corresponding to the cellular problem under consideration and establish the relation between the values of the Lagrange and Castigliano functionals for the cellular problem and the bending stiffness of a beam.

The Lagrange functional $J_u(\mathbf{u})$ for cellular problem (4.2.1) is the following:

$$J_u(\mathbf{u}) = (1/2)\langle 2y_\alpha c_{ij11}(\mathbf{y})u_{i,j} - c_{ijkl}(\mathbf{y})u_{i,j}u_{k,l}\rangle . \tag{4.2.8}$$

It is considered on the set of virtual displacements,

$$V = \{\mathbf{u} \in \{H^1(Y)\}^3 \ : \mathbf{u}(\mathbf{y}) \text{ is periodic in } y_1 \text{ with period } T\}. \tag{4.2.9}$$

Let us introduce (in formally, for the moment) the Castigliano functional as follows:

$$J_\sigma(\sigma) = (1/2)\langle c^{-1}_{ijkl}(\mathbf{y})\sigma_{ij}\sigma_{kl} + 2\sigma_{11}y_\alpha + y^2_\alpha c_{1111}(\mathbf{y})\rangle \tag{4.2.10}$$

$$= (1/2)\langle c^{-1}_{ijkl}(\mathbf{y})[\sigma_{ij} + y_\alpha c_{ij11}(\mathbf{y})][\sigma_{kl} + y_\alpha c_{kl11}(\mathbf{y})]\rangle ,$$

where c^{-1}_{ijkl} means the tensor inverse with respect to tensor c_{ijkl}.

It is considered on the set of admissible stresses,

$$\Sigma = \{\sigma_{ij}(\mathbf{y}) \in L^6(Y) : \sigma_{ij,j} = 0 \quad \text{in } Y, \ \ \sigma_{ij}n_j = 0 \quad \text{on } S, \tag{4.2.11}$$

$$\sigma_{ij}n_j \text{ is periodic in } y_1 \text{ with period } T\}.$$

We establish the relation between Lagrange and Castigliano functionals.

Proposition 4.6. *Under conditions C (Sect.* 1.2), *the following equalities hold*:

$$\max_{\mathbf{u}\in V} J_u(\mathbf{u}) = \min_{\sigma\in\Sigma} J_\sigma(\sigma) = J_u(\mathbf{X}^\alpha), \tag{4.2.12}$$

where $\mathbf{X}^\alpha$ *is the solution of problem* (4.2.1).

Proof. Under conditions *C*, the functional $J_u(\mathbf{u})$ is strongly convex (Ekeland and Team, 1976) on the set $\{\mathbf{u} \in V : \langle \mathbf{u} \rangle = 0, \ \langle u_\gamma y_\gamma s_\Gamma \rangle = 0\}$ (it is the orthogonal complementary space of the kernel of the cellular problem). Then, the problem,

$$J_u(\mathbf{u}) \to \max, \ \mathbf{u} \in V, \tag{4.2.13}$$

has a unique solution belonging to the set $\{\mathbf{u} \in V : \langle \mathbf{u} \rangle = 0, \ \langle u_\gamma y_\gamma s_\Gamma \rangle = 0\}$ and problem (4.2.1) is Euler's equation for problem (4.2.13). Then, $\mathbf{X}^\alpha$ is the solution of problem (4.2.12). The Castigliano functional (4.2.9) under conditions C is strongly convex (Ekeland and Temam, 1976). Then, the problem,

$$J_\sigma(\sigma) \to \min, \ \sigma_{ij} \in \Sigma, \tag{4.2.14}$$

has a unique solution, which satisfies the following equation:

$$\int_Y [c^{-1}_{ijkl}(\mathbf{y})\sigma_{kl} + y_\alpha c_{ij11}(\mathbf{y})]\eta_{ij}(\mathbf{y})d\mathbf{y} = 0 \quad \text{for any } \eta_{ij} \in \Sigma. \tag{4.2.15}$$

Let us consider the stresses corresponding to the solution of problem (4.2.1),

$$\sigma_{ij} = c_{ijkl}(\mathbf{y})X^\alpha_{k,l} - y_\alpha c_{ij11}(\mathbf{y}), \tag{4.2.16}$$

and verify that they are the solution of problem (4.2.13). For that, it is enough to verify that stresses (4.2.16) belong to Σ and satisfy (4.2.15). Stresses (4.2.16) belong to Σ because $\mathbf{X}^\alpha$ is the solution of problem (4.2.1).

Substituting (4.2.16) in (4.2.15) and integrating by parts, we obtain

$$\int_Y X^{\alpha}_{i,j}\eta_{ij}\mathbf{dy} = -\int_Y X^{\alpha}_i\eta_{ij,j}\mathbf{dy} + \int_S X^{\alpha}_i\eta_{ij}n_j\mathbf{dy} + \int_{\omega_+\cup\omega_-} X^{\alpha}_i\eta_{ij}\mathbf{dy}, \tag{4.2.17}$$

where ω_+ and ω_- denote the surfaces of domain Y perpendicular to the Oy_1 axis (Fig. 4.3).

The right-hand part of (4.2.17) is equal to zero by virtue of $\mathbf{X}^{\alpha} \in V$, $\eta_{ij} \in \Sigma$; see definition (4.2.11).

To complete the proof, one must verify that $J_u(\mathbf{X}^{\alpha}) = J_{\sigma}(c_{ijkl}(\mathbf{y})X^{\alpha}_{k,l} - y_{\alpha}c_{ij11}(\mathbf{y}))$ or (that is the same) that the following equality is valid:

$$\langle 2y_{\alpha}c_{ij11}(\mathbf{y})X^{\alpha}_{i,j} - c_{ijkl}(\mathbf{y})X^{\alpha}_{i,j}X^{\alpha}_{k,l}\rangle \tag{4.2.18}$$

$$= \langle c^{-1}_{ijkl}(\mathbf{y})[\sigma_{ij} + y_{\alpha}c_{ij11}(\mathbf{y})][\sigma_{kl} + y_{\alpha}c_{kl11}(\mathbf{y})]\rangle .$$

Substituting (4.2.16) in (4.2.18), we obtain the equality,

$$\langle 2y_{\alpha}c_{ij11}(\mathbf{y})X^{\alpha}_{i,j} - c_{ijkl}(\mathbf{y})X^{\alpha}_{i,j}X^{\alpha}_{k,l}\rangle = \langle c_{ijkl}(\mathbf{y})X^{\alpha}_{i,j}X^{\alpha}_{k,l}\rangle ,$$

which is true by virtue of (4.2.4).

We establish the relation between the Lagrange and Castigliano functionals and bending stiffnesses. From (4.2.3) and (4.2.8), we obtain

$$A^2_{\alpha\alpha} = \langle y^2_3 c_{1111}(\mathbf{y})\rangle - 2\max_{\mathbf{u}\in V} J_u(\mathbf{u}) \ \ (\alpha = 2, 3). \tag{4.2.19}$$

From (4.2.12) and (4.2.19), we obtain

$$A^2_{\alpha\alpha} = \langle y^2_3 c_{1111}(\mathbf{y})\rangle - 2\min_{\sigma\in\Sigma} J_{\sigma}(\sigma). \tag{4.2.20}$$

These are two variational principles (one in terms of displacements and the other in terms of stresses) for bending stiffnesses of a nonuniform nonhomogeneous beam.

When $\mathbf{u} \in V$ and $\sigma_{ij} \in \Sigma$ are arbitrary, we obtain from (4.2.20) the following two-sided estimates for bending stiffnesses:

$$\langle y^2_{\alpha} c_{1111}(\mathbf{y})\rangle - 2J_u(\mathbf{u}) \geq A^2_{\alpha\alpha} \geq \langle y^2_3 c_{1111}(\mathbf{y})\rangle - 2J_{\sigma}(\sigma). \tag{4.2.21}$$

Taking into account (4.2.10), one can rewrite the lower estimate in the following suitable form:

$$A_{\alpha\alpha}^{2} \geq \langle -c_{ijkl}^{-1}(\mathbf{y})\sigma_{ij}\sigma_{kl} - 2\sigma_{11}y_{\alpha}\rangle . \qquad (4.2.22)$$

Variational principles and estimates for torsional stiffness

The Lagrange functional (4.2.5) is the following:

$$J_{u}(\mathbf{u}) = (1/2)\langle 2y_{\Gamma}s_{\gamma}c_{ij\gamma 1}(\mathbf{y})u_{i,j} - c_{ijkl}(\mathbf{y})u_{i,j}u_{k,l}\rangle . \qquad (4.2.23)$$

It is considered on set V (4.2.9).

We introduce the Castigliano functional as follows:

$$J_{\sigma}(\sigma) = (1/2)\langle c_{ijkl}^{-1}(\mathbf{y})\sigma_{ij}\sigma_{kl} - 2y_{\Gamma}s_{\gamma}\sigma_{\gamma 1} + y_{\gamma}^{2}c_{1\gamma 1\gamma}(\mathbf{y})\rangle \qquad (4.2.24)$$

$$= (1/2)\langle c_{ijkl}^{-1}(\mathbf{y})[\sigma_{ij} - y_{\Gamma}s_{\gamma}c_{ij\gamma 1}(\mathbf{y})][\sigma_{kl} - y_{\Gamma}s_{\gamma}c_{kl\gamma 1}(\mathbf{y})]\rangle .$$

It is considered on set Σ (4.2.9).

Proposition 4.7. *Under conditions C, the following equations hold for the functionals determined by*

$$\max_{\mathbf{u}\in V} J_{u}(\mathbf{u}) = \min_{\sigma\in\Sigma} J_{\sigma}(\sigma) = J_{u}(\mathbf{X}) , \qquad (4.2.25)$$

where $\mathbf{X}(\mathbf{y})$ *is the solution of problem* (4.2.5).

Proof. The proof is analogous to the proof of Proposition 4.6. Under conditions C, the problem,

$$J_{\sigma}(\sigma) \to \min , \; \sigma_{ij} \in \Sigma , \qquad (4.2.26)$$

has a unique solution, which satisfies the following equation:

$$\int_{Y} [c_{ijkl}^{-1}(\mathbf{y})\sigma_{kl} - y_{\Gamma}s_{\gamma}c_{ij\gamma 1}(\mathbf{y})]\eta_{ij}(\mathbf{y})d\mathbf{y} = 0 \quad \text{for any } \eta_{ij} \in \Sigma . \qquad (4.2.27)$$

Let us consider the stresses corresponding to the solution of problem (4.2.5),

$$\sigma_{ij} = c_{ijkl}(\mathbf{y})X_{k,l} + y_{\Gamma}s_{\gamma}c_{ij\gamma 1}(\mathbf{y}) , \qquad (4.2.28)$$

and verify that they are the solution of problem (4.2.26). One can prove it was done in Proposition 4.6. To complete the proof, one must verify that

$J_u(\mathbf{X}) = J_\sigma(c_{ijkl}(\mathbf{y})X_{k,l} + y_\Gamma s_\gamma c_{ij\gamma 1}(\mathbf{y}))$. This follows by direct computation.

Let us establish the relation between Lagrange and Castigliano functionals and torsional stiffness. From (4.2.7) and (4.2.23), we obtain

$$B = \langle y_\gamma^2 c_{1\gamma 1\gamma}(\mathbf{y})\rangle - 2\max_{\mathbf{u}\in V} J_u(\mathbf{u}) . \tag{4.2.29}$$

From (4.2.25) and (4.2.29), we obtain the following equalities:

$$B = \langle y_\gamma^2 c_{1\gamma 1\gamma}(\mathbf{y})\rangle - 2\min_{\sigma\in\Sigma} J_\sigma(\sigma) . \tag{4.2.30}$$

These are two variational principles (one in terms of displacements and the other in terms of stresses) for the torsional stiffness of a beam of arbitrary structure. When $\mathbf{u}\in V$ and $\sigma_{ij}\in\Sigma$ are arbitrary, we obtain from (4.2.30) the following two-sided estimates for torsional stiffness:

$$\langle y_\gamma^2 c_{1\gamma 1\gamma}(\mathbf{y})\rangle - 2J_u(\mathbf{u}) \geq B \geq \langle y_\gamma^2 c_{1\gamma 1\gamma}(\mathbf{y})\rangle - 2J_\sigma(\sigma) . \tag{4.2.31}$$

The right-hand side inequality from (4.2.31) can be rewritten in the following form:

$$B \geq \langle -c_{ijkl}^{-1}(\mathbf{y})\sigma_{ij}\sigma_{kl} - 2y_\Gamma s_\gamma \sigma_{\gamma 1}\rangle . \tag{4.2.32}$$

Variational principle for shear stiffnesses $A_{23}^2 = A_{32}^2$

The technique used above to derive variational principles is suitable for symmetrical quadratic forms. Shear stiffnesses $A_{23}^2 = A_{32}^2$ are computed in accordance with the formulas (see Sect. 3.2),

$$A_{23}^2 = \langle y_2[y_3 c_{1111}(\mathbf{y}) - c_{ijkl}(\mathbf{y})\mathrm{X}_{k,l}^3(\mathbf{y})]\rangle , \tag{4.2.33}$$

$$A_{32}^2 = \langle y_3[y_2 c_{1111}(\mathbf{y}) - c_{ijkl}(\mathbf{y})\mathrm{X}_{k,l}^2(\mathbf{y})]\rangle , \tag{4.2.34}$$

which cannot be reduced to functionals symmetrical with respect to the coordinates y_2, y_3.

Note. It can be demonstrated that $A_{23}^2 = A_{32}^2$.

We introduce the following Lagrange functional:

$$J_u(\mathbf{u}) = (1/2)\langle 2(y_2 + y_3)c_{ij11}(\mathbf{y})u_{i,j} - c_{ijkl}(\mathbf{y})u_{i,j}u_{k,l}\rangle . \tag{4.2.35}$$

The solution of the problem,

$$J_u(\mathbf{u}) \to \max , \ \mathbf{u} \in V ,$$

is $\mathbf{X}^+ = \mathbf{X}^1 + \mathbf{X}^2$, where $\mathbf{X}^1$ and $\mathbf{X}^2$ are solutions of problem (4.2.1). Using this equality and the note above, we arrive at the equality,

$$A_{22}^2 + 2A_{23}^2 + A_{33}^2 = \langle c_{1111}(\mathbf{y})(y_2 + y_3)^2\rangle - 2\max_{\mathbf{u}\in V} J_u(\mathbf{u}) , \tag{4.2.36}$$

where $J_u(\mathbf{u})$ is given by (4.2.35).

Using (4.2.36), we obtain a pair of variational principles for the expression $A_{22}^2 + 2A_{23}^2 + A_{33}^2$. Then, as soon the bending stiffnesses A_{22}^2, A_{33}^2 are computed (or estimated), the stiffnesses $A_{23}^2 = A_{32}^2$ can be computed (estimated).

This case demonstrates that the variational principles for a cellular boundary value problem cannot be applied directly to computing of the shear stiffnesses of a beam.

Variational principles for axial stiffness

The cellular problem for computing axial stiffness A has the form (see Sect. 3.2),

$$\begin{cases} [c_{ijmn}(\mathbf{y})X_{m,ny} + c_{ij11}(\mathbf{y})]_{,j} = 0 \quad \text{in } Y, \\ [c_{ijmn}(\mathbf{y})X_{m,ny} + c_{ij11}(\mathbf{y})]n_j = 0 \quad \text{on } S, \\ \mathbf{X}(\mathbf{y}) \text{ is periodic in } y_1 \text{ with period } T. \end{cases} \tag{4.2.37}$$

Axial stiffness is computed in accordance with the formula,

$$A = \langle c_{1111}(\mathbf{y}) + c_{11kl}(\mathbf{y})X_{k,l}\rangle . \tag{4.2.38}$$

As above, variational principles and two-sided estimates can be derived. We present here only the results.

The Lagrange functional is the following:

$$J_u(\mathbf{u}) = (1/2)\langle 2c_{1111}(\mathbf{y})u_{i,j} - c_{ijkl}(\mathbf{y})u_{i,j}u_{k,l}\rangle . \qquad (4.2.39)$$

It is considered on set V (4.2.9).

The Castigliano functional is the following:

$$J_\sigma(\sigma) = (1/2)\langle c^{-1}_{ijkl}(\mathbf{y})\sigma_{ij}\sigma_{kl} + c_{1111}(\mathbf{y})\rangle . \qquad (4.2.40)$$

It is considered on set Σ (4.2.11).

Proposition 4.8. *Under conditions C, the following equalities hold:*

$$\max_{\mathbf{u}\in V} J_u(\mathbf{u}) = \min_{\sigma\in\Sigma} J_\sigma(\sigma) = J_u(\mathbf{X}), \qquad (4.2.41)$$

where $\mathbf{X}$ *is the solution of problem* (4.2.37).

Using Proposition 4.7, we obtain the following equalities:

$$A = \langle c_{1111}(\mathbf{y})\rangle - 2\max_{\mathbf{u}\in V} J_u(\mathbf{u}), \qquad (4.2.42)$$

$$A = \langle c_{1111}(\mathbf{y})\rangle - 2\min_{\sigma\in\Sigma} J_\sigma(\sigma).$$

These are two variational principles (one in terms of displacements and the other in terms of stresses) for the axial stiffnesses of a beam of arbitrary structure. When $\mathbf{u}\in V$ and $\sigma_{ij}\in\Sigma$ are arbitrary, we obtain from (4.2.42) the following two-sided estimates:

$$\langle c_{1111}(\mathbf{y})\rangle - 2J_u(\mathbf{u}) \geq A \geq \langle c_{1111}(\mathbf{y})\rangle - J_\sigma(\sigma). \qquad (4.2.43)$$

Variational principles for coupling stiffness

For a beam of arbitrary nonhomogeneous structure, there is no *a priori* way to introduce centers of bending and torsion. A similar problem was discussed by Washizu, (1982, appendix G) and Kolpakov (1998b) in connection with the "neutral" axis of a nonhomogeneous beam. As a result, one cannot decouple the axial strain from bending and torsion, and coupling stiffnesses A^1_α must be considered. Coupling stiffnesses A^1_α are given by the formula (see Sect. 3.2),

$$A^1_\alpha = \langle y_\alpha c_{1111}(\mathbf{y}) - c_{11kl}(\mathbf{y})X^\alpha_{k,l}(\mathbf{y})\rangle = \langle y_\alpha[c_{1111}(\mathbf{y}) - c_{11kl}(\mathbf{y})X_{k,l}(\mathbf{y})]\rangle , \qquad (4.2.44)$$

where $\mathbf{X}^{\alpha}$ is the solution of problem (4.2.1) and $\mathbf{X}$ is the solution of problem (4.2.37).

In order to derive the variational principle for coupling stiffnesses, we consider the following functional:

$$J_u(\mathbf{u}) = (1/2)\langle 2(y_\alpha + h)c_{ij11}(\mathbf{y})u_{i,j} - c_{ijkl}(\mathbf{y})u_{i,j}u_{k,l}\rangle, \qquad (4.2.45)$$

where h is an arbitrary nonzero real number.

The solution of the problem,

$$J_u(\mathbf{u}) \to \max, \qquad \mathbf{u} \in V,$$

is $\mathbf{X}^{+} = \mathbf{X}^{\alpha} + h\mathbf{X}$. Using this equality, we can verify that

$$A^2_{\alpha\alpha} + 2hA^1_{\alpha} + h^2 A^0_{11} = \langle c_{1111}(\mathbf{y})\rangle h^2 - 2J_u(\mathbf{X}^{+}), \qquad (4.2.46)$$

where $J_u(\mathbf{u})$ is given by (4.2.45).

Using (4.2.46), we obtain a pair of variational principles and two-sided estimates for the expression, $A^2_{\alpha\alpha} + 2hA^1_{\alpha} + h^2 A^0_{11}$. Using these principles and estimates, we can compute (or estimate) the coupling stiffnesses if the bending and axial stiffnesses are computed (estimated).

Some estimates for the stiffnesses of a nonhomogeneous beam

We present here some estimates following from the variational principles obtained above.

The Voight-like upper bound for stiffnesses

Putting $\mathbf{u} = 0$ into (4.2.21), (4.2.31), and (4.2.43) (it is clear that $0 \in V$), we obtain the following estimate for stiffnesses

$$A^2_{\alpha\alpha} \le \langle y_\alpha^2 c_{1111}(\mathbf{y})\rangle = \left\langle \frac{[1-\nu(\mathbf{y})]E(\mathbf{y})y_\alpha^2}{[1+\nu(\mathbf{y})][1-2\nu(\mathbf{y})]} \right\rangle, \qquad (4.2.47)$$

$$A^2_{22} + 2A^2_{32} + A^2_{33} \le \langle c_{1111}(\mathbf{y})(y_2 + y_3)^2\rangle = \left\langle \frac{[1-\nu(\mathbf{y})]E(\mathbf{y})(y_2 + y_3)^2}{[1+\nu(\mathbf{y})][1-2\nu(\mathbf{y})]} \right\rangle;$$

$$B \le \langle y_\gamma^2 c_{1\gamma 1\gamma}(\mathbf{y})\rangle = \left\langle \frac{E(\mathbf{y})(y_2^2 + y_3^2)}{1+\nu(\mathbf{y})} \right\rangle , \tag{4.2.48}$$

$$A \le \langle c_{1111}(\mathbf{y})\rangle = \left\langle \frac{E(\mathbf{y})[1-\nu(\mathbf{y})]}{[1+\nu(\mathbf{y})][1-2\nu(\mathbf{y})]} \right\rangle ;$$

$$A_{\alpha\alpha}^2 + 2hA_\alpha^1 + h^2 A_{11}^0 \le \langle c_{1111}(\mathbf{y})\rangle h^2 = \left\langle \frac{E(\mathbf{y})[1-\nu(\mathbf{y})]}{[1+\nu(\mathbf{y})][1-2\nu(\mathbf{y})]} \right\rangle h^2 . \tag{4.2.49}$$

The last equations in (4.2.47)–(4.2.49) are written for a beam made of isotropic materials: $E, \nu = E, \nu(\mathbf{y})$ are Young's modulus and Poisson's ratio for the materials from which the beam is made.

Estimates (4.2.47), (4.2.48) can be considered nondirect beam analogies of Voight's bound (Voight, 1928).

Cylindrical beam. Lower estimates for bending stiffnesses

Let the periodicity cell be a cylinder, $Y = S \times [0,1]$, $S \subset R^2$. The distribution of the material characteristics over the beam may be arbitrary.

Consider the stress tensor of the form, $\sigma_{11} = Cy_\alpha^n$, where C and n are arbitrary real numbers and $\sigma_{ij} = 0$ if $ij \ne 11$. It is obvious that $\sigma_{ij} \in \Sigma$. For these stresses, the expression on the right-hand side of (4.2.22) takes the form,

$$-\langle c_{1111}^{-1}(\mathbf{y}) y_\alpha^{2n}\rangle C^2 - 2\langle y_\alpha^{n+1}\rangle C . \tag{4.2.50}$$

There are no restrictions on C. Then, one can maximize (4.2.50) with respect to C. Euler's equation for (4.2.50) is

$$-\langle c_{1111}^{-1}(\mathbf{y}) y_\alpha^{2n}\rangle C - 2\langle y_\alpha^{n+1}\rangle = 0 .$$

Its solution is $C = -\dfrac{2\langle y_\alpha^{n+1}\rangle}{\langle c_{1111}^{-1}(\mathbf{y}) y_\alpha^{2n}\rangle}$. Substituting this solution in (4.2.50), we obtain the following estimate:

$$A_{\alpha\alpha}^2 \ge \frac{\langle y_\alpha^{n+1}\rangle^2}{\langle c_{1111}^{-1}(\mathbf{y}) y_\alpha^{2n}\rangle} . \tag{4.2.51}$$

If the material of the beam is isotropic, then (4.2.51) takes the form,

$$A_{\alpha\alpha}^2 \geq \frac{\langle y_\alpha^{n+1} \rangle^2}{\left\langle \dfrac{y_\alpha^{2n}}{E(\mathbf{y})} \right\rangle}. \tag{4.2.52}$$

When $n = 0$ (the case of uniform admissible stresses), we obtain the following beam analogue of Reuss's bound:

$$A_{\alpha\alpha}^2 \geq \frac{\langle y_\alpha \rangle^2}{\left\langle \dfrac{1}{E(\mathbf{y})} \right\rangle}.$$

For a symmetrical beam, $\langle y_\alpha \rangle^2 = 0$, and the estimate takes the trivial form, $A_{\alpha\alpha}^2 \geq 0$. The first nontrivial case takes place if $n = 1$. In this case, we obtain the following estimate:

$$A_{\alpha\alpha}^2 \geq \frac{\langle y_\alpha^2 \rangle^2}{\left\langle \dfrac{y_\alpha^2}{E(\mathbf{y})} \right\rangle}.$$

Cylindrical beam of coaxial structure

Let periodicity cell Y be a cylinder and, in addition, the elastic constants depend only on the spatial variables y_2, y_3. A beam of this kind is called a coaxial beam structure. Let the coaxial beam be made of isotropic materials. We consider the stress tensor of the form, $\sigma_{11} = \sigma_{11}(y_2, y_3)$, and $\sigma_{ij} = 0$ if $ij \neq 11$. It is obvious that $\sigma_{ij} \in \Sigma$. In the case under consideration, the right-hand side of (4.2.22) takes the form,

$$-\left\langle \frac{\sigma_{11}^2}{E(y_2, y_3)} - 2\sigma_{11} y_\alpha \right\rangle. \tag{4.2.53}$$

There are no restrictions on σ_{11}. Then, one can maximize (4.2.53) with respect to σ_{11}. Euler's equation for (4.2.53) has the form

$$-\frac{\sigma_{11}}{E(\mathbf{y})} - y_\alpha = 0. \tag{4.2.54}$$

The solution of (4.2.54) is $\sigma_{11} = -E(y_2, y_3) y_\alpha$. Substituting this solution in

(4.2.53), we obtain the following estimate:

$$A_{\alpha\alpha}^2 \geq \langle y_\alpha^2 E(y_2, y_3) \rangle . \tag{4.2.55}$$

It is of interest that the Voight-type estimate is derived from the Castigliano variational principle and gives the lower bound.

Circular cylindrical beam. Lower bound for torsional stiffness

Let the periodicity cell be a cylinder $Y = S \times [0,1]$, where S is a circle. Consider a stress tensor of the form, $\sigma_{12} = Cy_3$, $\sigma_{13} = -Cy_2$, where C is an arbitrary constant, and $\sigma_{ij} = 0$ if $ij \neq 12, 21, 13, 31$. It is obvious that $\sigma_{ij} \in \Sigma$ if the cylinder is circular. For these stresses, the right-hand side of (4.2.32) takes the form,

$$-\langle c_{1\gamma 1\gamma}^{-1}(\mathbf{y})(y_2^2 + y_3^2) \rangle C^2 - 2\langle y_2^2 + y_3^2 \rangle C . \tag{4.2.56}$$

There are no restrictions on C. Then, one can maximize (4.2.56) with respect to C. Euler's equation for (4.2.56) is

$$-\langle c_{1\gamma 1\gamma}^{-1}(\mathbf{y})(y_2^2 + y_3^2) \rangle C - \langle y_2^2 + y_3^2 \rangle = 0 .$$

Its solution is

$$C = -\frac{\langle y_2^2 + y_3^2 \rangle}{\langle c_{1\gamma 1\gamma}^{-1}(\mathbf{y})(y_2^2 + y_3^2) \rangle} . \tag{4.2.57}$$

Substituting (4.2.57) in (4.2.56), we obtain

$$A \geq \frac{\langle y_2^2 + y_3^2 \rangle^2}{\langle c_{1\gamma 1\gamma}^{-1}(\mathbf{y})(y_2^2 + y_3^2) \rangle} .$$

If the beam is made of an isotropic material, then (4.2.32) takes the form,

$$A \geq \frac{\langle y_2^2 + y_3^2 \rangle^2}{\left\langle \dfrac{[y_2^2 + y_3^2][1 + \nu(\mathbf{y})]}{E(\mathbf{y})} \right\rangle} . \tag{4.2.58}$$

Note that (4.2.58) is not a Reuss's bound.

4.3 The Homogenized Method for Lattice Plates

The problem of describing lattice plates (net and framework plates and so on) with continuous models, as well as the inverse problem approximation of continuum structures with framework models (see, e.g., Hrennikoff, 1941) is a problem of permanent interest in structural mechanics, and it has been raised in the literature owing to the development of theoretical and computational methods (Noor, 1988; Pshenichnov, 1982).

The problem can be effectively solved with the homogenization method. It is possible to develop an engineering theory of homogenization for periodic frameworks combining the homogenization method with the classical methods of the theory of strength. Such a theory was developed in the book by Annin et al. (1993) for structures made of plates. In the following sections, a version of the homogenization procedure is developed for lattice plates and beams.

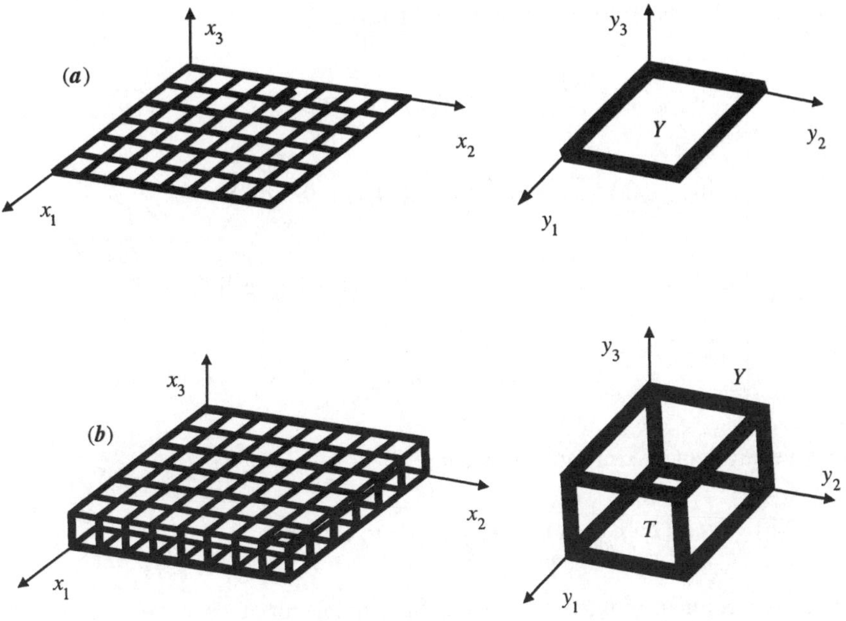

Fig. 4.4. An overview and periodicity cells of lattice plates: (***a***) the network plate and (***b***) lattice plate.

We call a thin periodic structure formed by beams or rods a lattice plate. A lattice plate looks like a solid plate from the overall point of view. Lattice plates are a generalization of network plates studied by Pshenichnov (1982). The difference between the network and the lattice plate is the structure across the thickness. The network plate has a complex structure in that plane, and simple (one-level) structure across the thickness; Fig. 4.4(***a***). The lattice plate has a complex (multilevel) structure both in that plane and across that thickness; Fig. 4.4(***b***). Both network and lattice plates have found numerous applications in many branches of modern engineering.

Derivation of the global and local equations of equilibrium and calculation of the stiffnesses for network plates took great effort in the book mentioned (Pshenichnov, 1982). The variational principles presented in Sect. 4.1 give an effective and accurate method for calculating the overall characteristics of both network and lattice plates.

Transformation of the cellular boundary-value problem into a finite-dimensional problem for lattice plate-like structures

The correspondence between the local characteristics of an inhomogeneous structure and its homogenized characteristics is established as follows (see Sect. 4.1). We solve the following cellular problem:

$$\begin{cases} [c_{ijmn}(\mathbf{y})X^{\nu\alpha\beta}_{m,n} + (-1)^{\nu} y_3^{\nu} c_{ij\beta\alpha}(\mathbf{y})]_{,j} = 0 \quad \text{in } Y, \\ [c_{ijmn}(\mathbf{y})X^{\nu\alpha\beta}_{m,n} + (-1)^{\nu} y_3^{\nu} c_{ij\beta\alpha}(\mathbf{y})]n_j = 0 \quad \text{on } S, \\ \mathbf{X}^{\nu\alpha\beta}(\mathbf{y}) \text{ is periodic in } y_1, y_2 \text{ with periodicity cell } T, \\ \langle \mathbf{X}^{\nu\alpha\beta} \rangle = 0, \end{cases} \tag{4.3.1}$$

or the equivalent maximization problem,

$$J_u(\mathbf{u}) = (1/2)\langle 2y_3^{\nu} c_{ij\alpha\beta}(\mathbf{y})u_{i,j} - c_{ijkl}(\mathbf{y})u_{i,j}u_{k,l}\rangle \to \max, \qquad \mathbf{u} \in V.$$

After the cellular problem is solved, the homogenized characteristics can be computed in accordance with the formula [see (4.1.3)]:

$$A^{\nu+\mu}_{\gamma\delta\alpha\beta} = \tag{4.3.2}$$

$$\langle c_{ijkl}(\mathbf{y})[X^{\nu\gamma\delta}_{k,l}(\mathbf{y})+(-1)^{\nu}y_3^{\nu}\delta_{k\gamma}\delta_{l\delta}][X^{\nu\alpha\beta}_{i,j}(\mathbf{y})+(-1)^{\mu}y_3^{\mu}\delta_{i\alpha}\delta_{j\beta}]\rangle,$$

ν,μ take values 0, 1; $\alpha,\beta,\gamma,\delta=1,2$. Here, $A^0_{\gamma\delta\alpha\beta}$, $A^1_{\gamma\delta\alpha\beta}$, $A^2_{\gamma\delta\alpha\beta}$ mean the in-plane, coupling, and bending stiffnesses of the homogenized plate.

Transformation of the cellular problem

Problem (4.3.1) represents elastic problems of a special form. The specifics of the problems are related to the free terms and boundary conditions. The boundary conditions cannot be transformed into boundary conditions of another type (except for a symmetrical periodicity cell). However, this does not raise a problem because the periodicity conditions can be easily reformulated in terms of the theory of strength.

In solving the cellular problem, the difficulties arise from the terms, $[y_3^{\nu}c_{ij\beta\alpha}(\mathbf{y})]_{,j}$, which can be treated as no typical mass forces. We demonstrate that this term can be eliminated from cellular problems (4.3.1). For $\nu=0$, we can do it by a known change of function. We introduce

$$\mathbf{W}^{0\alpha\beta}(\mathbf{y})=\mathbf{X}^{0\alpha\beta}(\mathbf{y})+y_{\alpha}\mathbf{e}_{\beta}\ . \tag{4.3.3}$$

After that, (4.3.1) with respect to the function $\mathbf{W}(\mathbf{y})$ take the form of the usual equations of elasticity theory with no mass forces.

Let us demonstrate that a similar procedure can be used when $\nu=1$ (for problems corresponding to bending and torsion). The problem is equivalent to the problem of the existence of displacement $\xi^{0\alpha\beta}(\mathbf{y})$ such that for strains $e_{kl}(\xi)=(1/2)(\partial\xi_k/\partial y_l+\partial\xi_l/\partial y_k)$, the following conditions are satisfied:

$$c_{ijmn}(\mathbf{y})e_{mn}-y_3c_{ij\beta\alpha}(\mathbf{y})=0\ . \tag{4.3.4}$$

Solving (4.3.4) with respect to $\{e_{mn}\}$, we obtain

$$e_{mn}=-c^{-1}_{mnij}(\mathbf{y})c_{ij\beta\alpha}(\mathbf{y})y_3\,, \tag{4.3.5}$$

where $c^{-1}_{mnij}(\mathbf{y})$ means the tensor inverse with respect to tensor $c_{mnij}(\mathbf{y})$.

It follows from (4.3.5) that

$$e_{mn}=-\delta_{m\alpha}\delta_{n\beta}y_3\,. \tag{4.3.6}$$

The equations of compatibility for strains have the following form (Love, 1929):

$$e_{ik,js} - e_{kj,is} - e_{is,jk} + e_{sj,ik} = 0 . \tag{4.3.7}$$

The compatibility equations (4.3.7) are satisfied by the strains (4.3.6), all of which represent linear functions of variables y_1, y_2, y_3. Thus, displacements $\xi^{0\alpha\beta}(\mathbf{y})$, satisfying (4.3.4), exist.

Computation of function $\xi^{\nu\alpha\beta}(\mathbf{y})$

Note that the indexes α, β take values 1 and 2, when cellular problems for a plate are considered.

Computation of function $\xi^{111}(\mathbf{y})$ ($\alpha\beta$ =11)

In this case the strains are the following:

$$e_{11} = y_3 , \; e_{mn} = 0 \quad \text{if} \quad mn \neq 11 . \tag{4.3.8}$$

We write equalities (4.3.8) for strains e_{ii}:

$$e_{11} = \frac{\partial u_1}{\partial y_1} = y_3 , \; e_{22} = \frac{\partial u_2}{\partial y_2} = 0 , \; e_{33} = \frac{\partial u_3}{\partial y_3} = 0 . \tag{4.3.9}$$

From (4.3.9), we obtain

$$\xi_1 = y_1 y_3 + f(y_2, y_3) , \; \xi_2 = p(y_1, y_3) , \; \xi_3 = p(y_1, y_2) . \tag{4.3.10}$$

From (4.3.8) and (4.3.10),

$$2e_{13} = y_1 + \frac{\partial f}{\partial y_3}(y_2, y_3) + \frac{\partial q}{\partial y_1}(y_1, y_2) = 0 , \tag{4.3.11}$$

$$2e_{12} = \frac{\partial f}{\partial y_3}(y_2, y_3) + \frac{\partial p}{\partial y_1}(y_1, y_3) = 0 ,$$

$$2e_{23} = \frac{\partial p}{\partial y_3}(y_1, y_3) + \frac{\partial q}{\partial y_2}(y_1, y_2) = 0 .$$

From the first equation in (4.3.11), it follows that

$$q = \frac{-y_1^2}{2} + Q(y_1) . \tag{4.3.12}$$

From the third equation in (4.3.11), it follows that

$$p = t(y_1)y_3 + U(y_1), \quad q = -t(y_1)y_2 + V(y_1). \tag{4.3.13}$$

From the first two equations in (4.3.11) and equations (4.3.12), (4.3.13), it follows that

$$y_1 + \frac{\partial f}{\partial y_3}(y_2, y_3) - \frac{\partial t}{\partial y_1}(y_1)y_2 + \frac{\partial V}{\partial y_1}(y_1) = 0, \tag{4.3.14}$$

$$\frac{\partial f}{\partial y_3}(y_2, y_3) + \frac{\partial t}{\partial y_1}(y_1)y_3 + \frac{\partial U}{\partial y_1}(y_1) = 0.$$

Equation (4.3.14) will be satisfied by

$$\frac{\partial t}{\partial y_1}(y_1) = 0, \quad f = 0, \quad \frac{\partial U}{\partial y_1}(y_1) = 0, \quad \frac{\partial V}{\partial y_1}(y_1) = -y_1.$$

As a result, we obtain

$$\xi_1^{111}(\mathbf{y}) = y_1 y_3, \tag{4.3.15}$$

$$\xi_2^{111}(\mathbf{y}) = 0,$$

$$\xi_3^{111}(\mathbf{y}) = \frac{-y_1^2}{2}.$$

Computation of function $\xi^{122}(\mathbf{y})$ ($\alpha\beta$ =22)

Function $\xi^{122}(\mathbf{y})$ can be obtained by replacing index $\alpha\beta$ =11 with $\alpha\beta$ =22. Thus,

$$\xi_1^{122}(\mathbf{y}) = 0, \tag{4.3.16}$$

$$\xi_2^{122}(\mathbf{y}) = y_2 y_3,$$

$$\xi_3^{122}(\mathbf{y}) = \frac{-y_2^2}{2}.$$

Computation of function $\xi^{112}(\mathbf{y})$ ($\alpha\beta$ =12)

In this case, the strains are the following:

$$e_{12} = y_3\,,\ e_{mn} = 0 \quad \text{if} \quad mn \neq 12\,. \tag{4.3.17}$$

We write equalities (4.3.17) for e_{ii}:

$$e_{11} = \frac{\partial \xi_1}{\partial y_1} = 0\,,\ e_{22} = \frac{\partial \xi_2}{\partial y_2} = 0\,,\ e_{33} = \frac{\partial \xi_3}{\partial y_3} = 0\,. \tag{4.3.18}$$

From (4.3.18),

$$\xi_1 = f(y_2, y_3)\,,\ \xi_2 = p(y_1, y_3)\,,\ \xi_3 = p(y_1, y_2)\,. \tag{4.3.19}$$

From (4.3.17) and (4.3.19),

$$2e_{13} = \frac{\partial f}{\partial y_3}(y_2, y_3) + \frac{\partial q}{\partial y_1}(y_1, y_2) = 0\,, \tag{4.3.20}$$

$$2e_{12} = \frac{\partial f}{\partial y_3}(y_2, y_3) + \frac{\partial p}{\partial y_1}(y_1, y_3) = 2y_3\,,$$

$$2e_{23} = \frac{\partial p}{\partial y_3}(y_1, y_3) + \frac{\partial q}{\partial y_2}(y_1, y_2) = 0\,.$$

From the first equation in (4.3.20), it follows that

$$f = s(y_2)y_3 + U(y_2)\,,\ q = -s(y_2)y_1 + V(y_2)\,. \tag{4.3.21}$$

From the third equation in (4.3.18), it follows that

$$p = t(y_1)y_3 + u(y_1)\,,\ q = -t(y_1)y_2 + v(y_1)\,. \tag{4.3.22}$$

Then,

$$q = -y_1 y_2\,,\ s = y_2\,,\ t = -y_1\,. \tag{4.3.23}$$

From the third equation in (4.3.20) and (4.3.22),

$$\frac{\partial s}{\partial y_2}(y_2)y_3 + \frac{\partial t}{\partial y_1}(y_1)y_3 = 2y_3\,. \tag{4.3.24}$$

From (4.3.23) and (4.3.24),

$$f = y_2 y_3,\ q = -y_1 y_2,\ p = -y_1. \tag{4.3.25}$$

As a result, we obtain

$$\xi_1^{112}(\mathbf{y}) = y_2 y_3, \tag{4.3.26}$$

$$\xi_2^{112}(\mathbf{y}) = y_1 y_3,$$

$$\xi_3^{112}(\mathbf{y}) = -y_1 y_2.$$

Computation of function $\xi^{21}(\mathbf{y})$ ($\alpha\beta$=21)
The function can be obtained by replacing the index $\alpha\beta$=12 with $\alpha\beta$=21. This replacement leads to (4.3.26).

The function $\xi^{0\alpha\beta}(\mathbf{y}) = y_\alpha \mathbf{e}_\beta$. Thus, we constructed function $\xi^{\nu\alpha\beta}(\mathbf{y})$ for all possible values of the indexes. Introducing functions

$$\mathbf{W}^{\nu\alpha\beta}(\mathbf{y}) = \mathbf{X}^{\nu\alpha\beta}(\mathbf{y}) + \xi^{\nu\alpha\beta}(\mathbf{y}), \tag{4.3.27}$$

we can write cellular problem (4.3.1) in the form

$$\begin{cases} [c_{ijmn}(\mathbf{y}) W_{m,ny}]_{,j} = 0 \quad \text{in } Y, \\ c_{ijmn}(\mathbf{y}) W_{m,ny} n_j = 0 \quad \text{on } S, \\ \mathbf{W}(\mathbf{y}) \text{ is periodic in } y_1, y_2 \text{ with periodicity cell } T, \\ \langle \mathbf{W} - \xi^{\nu\alpha\beta}(\mathbf{y}) \rangle = 0. \end{cases} \tag{4.3.28}$$

Problem (4.3.28) has the same form for all types of cellular problems [for all values of indexes (ν, α, β)]. Indexes (ν, α, β) come to the problem with function $\xi^{\nu\alpha\beta}(\mathbf{y})$. Functions $\xi^{\nu\alpha\beta}(\mathbf{y})$ correspond to the basic global deformations of a plate.

Method solution of the cellular problem (4.3.28).

Problem (4.3.28) is a problem of elasticity theory. We consider lattice plates, whose periodicity cell is made of beams/rods. Thus, periodicity cell problem (4.3.28) can be transformed into a periodicity problem of beam/rod theory.

The derivation of the cellular finite-dimensional problem meets some difficulties if one uses the approach of the theory of strength. The main one of them is related to the derivation of equilibrium and kinematic conditions in joint nodes. This problem can be solved by careful investigation of the conditions in joint nodes. Examples of such investigations can be found in the book by Pshenichnov (1982) for network plates and in the book by Kalamkarov and Kolpakov (1997) for spatial frameworks.

Using the variational principles presented in Sect. 4.1, we can derive a finite-dimensional cell problem for lattice plates of periodic structure in a clear and rigorous way. We will accept the classical kinematics hypotheses with a suitable hypotheses at joint nodes. It will define a set of admissible displacements. After that, we minimize the Lagrange functional $J_u(\mathbf{u})$ on the set of admissible displacements V_{disc}; see Sect. 4.1. If we accept the condition of rigid joints, we will arrive at the problem with the following conditions:

1. The forces applied to the elements forming the cellular structure are equal to zero.
2. The rigid-joint conditions are satisfied at the inner nodal points of the cellular structure, and the sum of the forces and sum of the moments at the nodal joints are equal to zero (in other words, the equilibrium conditions are satisfied for the inner nodes).
3. The conditions indicated in item 2 are satisfied at the nodal points of the cellular structure corresponding to the free surface S; see (4.3.1).
4. At boundary nodal joints, the periodicity conditions are satisfied for $\mathbf{W}^{\nu\alpha\beta}(\mathbf{y}) - \xi^{\nu\alpha\beta}(\mathbf{y})$ and for forces and moments.

Conditions 1–4 determine a finite-dimensional problem corresponding to cellular problem (4.3.1). Following the well-known procedure (see, e.g., Langhaar, 1962; Pipes, 1963; and Haug et al., 1986), we construct the finite-dimensional extremal principle, corresponding to the cellular problem,

$$J_u(\mathbf{u}) \to \max, \quad \mathbf{u} \in V_{disc}.$$

The behavior of beams and rods under the condition that the forces are equal to zero (condition 1) can be described in terms of generalized displacements of their ends. The connection conditions (conditions 2,3) will be satisfied by introducing generalized displacements automatically. The periodicity conditions (condition 3) also can be formulated in terms of generalized displacements of nodes:

$$\mathbf{u}_i = \mathbf{u}_{\Gamma(i)} + (\Xi_i^{\nu\alpha\beta} - \Xi_{\Gamma(i)}^{\nu\alpha\beta}), \tag{4.3.29}$$

where index i corresponds to a joint node that belong to the periodicity cell edge; $\Gamma(i)$ is the index of the corresponding node belonging to the opposite edge; $\{\mathbf{u}_i\}$ are the generalized displacements, which correspond to $\mathbf{W}(\mathbf{y})$; and $\Xi_i^{\nu\alpha\beta}$ mean the generalized displacements, which correspond to function $\xi^{\nu\alpha\beta}(\mathbf{y})$.

For nodes mentioned in conditions 2 and 3, forces and moments have the form,

$$\sum_{j\in K_i}[\mathbf{N}^{\nu\alpha\beta}(\mathbf{u}_i,\mathbf{u}_j)+\mathbf{Q}^{\nu\alpha\beta}(\mathbf{u}_i,\mathbf{u}_j)]=0, \tag{4.3.30}$$

$$\sum_{j\in K_i} M_\gamma^{\nu\alpha\beta}(\mathbf{u}_i,\mathbf{u}_j)=0,$$

where $\mathrm{K}(i)$ are indexes of nodes joined with node i by a structural element, $\mathbf{N}^{\nu\alpha\beta}(\mathbf{u}_i,\mathbf{u}_j)$ and $\mathbf{Q}^{\nu\alpha\beta}(\mathbf{u}_i,\mathbf{u}_j)$ are axial and shearing forces, and $M_\gamma^{\nu\alpha\beta}(\mathbf{u}_i,\mathbf{u}_j)$ are moments in the beam with the ends in the i th and j th nodes.

If the index i corresponds to a node lying at the periodicity cell boundary, we determine set $\mathrm{K}(i)$ by condition,

$$\mathrm{K}_i=\mathrm{K}_{\Gamma(i)}. \tag{4.3.31}$$

Then, for forces and moments at the boundary points, periodicity condition 4 takes the form (4.3.30)

Displacements $\mathbf{W}^{\nu\alpha\beta}(\mathbf{y})$ are determined from (4.3.29)-(4.3.31) to within the displacement of a solid. Then, condition $\langle\mathbf{W}^{\nu\alpha\beta}-\xi^{\nu\alpha\beta}(\mathbf{y})\rangle=0$ can be replaced by condition $\langle\mathbf{W}^{\nu\alpha\beta}\rangle=0$. The later condition can be written as

$$\sum_{i=1}^{m}\mathbf{u}_i=0, \tag{4.3.32}$$

where m is the total number of nodes.

Equations (4.3.29)-(4.3.32) represent an algebraic system of the form,

$$\mathbf{T}_b\{\mathbf{u}_i\}=\mathbf{b}^{\nu\alpha\beta}, \tag{4.3.33}$$

where $\mathbf{T}_b$ is a matrix, and $\mathbf{b}^{\nu\alpha\beta}$ is a vector.

System (4.3.33) is the cellular problem written in terms of the theory of material strength (and will be called the finite-dimensional cellular problem).

The procedure for deriving the finite-dimensional cellular problem is similar to the method described for frames in the book by Washizu (1982). The difference is related to the original variational principles used. We derive the finite-dimensional cellular problem from variational principles for the homogenized stiffnesses of the composite plate presented in Sect. 4.1.

Computation of homogenized stiffnesses of lattice plates

Formula (4.3.2) can be written as

$$A^{\nu+\mu}_{\gamma\delta\alpha\beta} = \langle c_{ijkl}(\mathbf{y})[X^{\nu\gamma\delta}_{k,l}(\mathbf{y}) + \xi^{\nu\gamma\delta}_{k,l}(\mathbf{y})][(X^{\mu\alpha\beta}_{i,j}(\mathbf{y}) + \xi^{\mu\alpha\beta}_{i,j}(\mathbf{y})]\rangle , \tag{4.3.34}$$

where $\xi^{\mu\gamma\delta}(\mathbf{y})$ is the function satisfying the condition

$$\xi^{\nu\gamma\delta}_{k,l}(\mathbf{y}) = (-1)^{\nu} y_3^{\nu} \delta_{i\alpha} \delta_{j\beta} \quad (\nu = 0,1 ;\ \alpha, \beta = 1,2). \tag{4.3.35}$$

Comparing (4.3.35) and (4.3.3), (4.3.8), we note that $\xi^{\mu\gamma\delta}(\mathbf{y})$ determined by (4.3.35) and the function determined by (4.3.3) is the same functions.

Using the notations (4.3.3), (4.3.27), we can write (4.3.34) as

$$A^{\nu+\mu}_{\gamma\delta\alpha\beta} = \langle c_{ijkl}(\mathbf{y}) W^{\nu\gamma\delta}_{k,l}(\mathbf{y}) W^{\mu\alpha\beta}_{i,j} \rangle , \tag{4.3.36}$$

where $\mathbf{W}^{\nu\alpha\beta}(\mathbf{y})$ is the solution of cellular problem (4.3.28).

The right-hand part of (4.3.36) can be written as the sum over the elements forming the cell:

$$A^{\nu+\mu}_{\gamma\delta\alpha\beta} = (mes T)^{-1} \int\limits_{Y} c_{ijkl}(\mathbf{y}) W^{\nu\gamma\delta}_{k,l}(\mathbf{y}) W^{\mu\alpha\beta}_{i,j} \mathbf{dy} \tag{4.3.37}$$

$$= (mes T)^{-1} \sum_{p=1}^{P} \int\limits_{L_p} c_{ijkl}(\mathbf{y}) W^{\nu\gamma\delta}_{k,l}(\mathbf{y}) W^{\mu\alpha\beta}_{i,j} \mathbf{dy} ,$$

where L_p means the pth beam forming the periodicity cell ($p = 1,...,P$), and P is the total number of beams forming the cell.

Let the beam L_p have the ends at the ith and jth nodes. The generalized displacements of the ends of the beam are $\mathbf{u}_i$ and $\mathbf{u}_j$. The integral,

$$\frac{1}{2} \sum_{p=1}^{P} \int\limits_{L_p} c_{ijkl}(\mathbf{y}) W^{\nu\gamma\delta}_{k,l}(\mathbf{y}) W^{\mu\alpha\beta}_{i,j} \mathbf{dy} ,$$

is equal to the elastic energy $E(\mathbf{u}_i, \mathbf{u}_j)$ of the beam L_p subjected to the generalized displacements $\mathbf{u}_i$ and $\mathbf{u}_j$. Thus,

$$A_{\gamma\delta\alpha\beta}^{\nu+\mu} = \frac{2}{mesT} \sum_{p=1}^{P} E(\mathbf{u}_i, \mathbf{u}_j). \tag{4.3.38}$$

The generalized displacements are determined from the solution of problem (4.3.33).

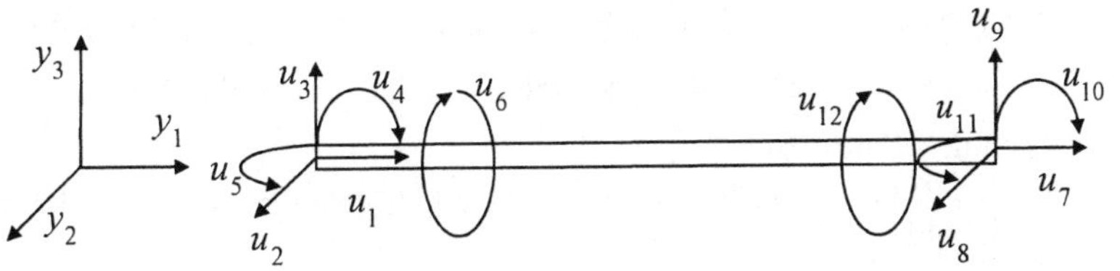

Fig. 4.5. The generalized displacements of a beam

We write the elastic energy $E(\mathbf{u}_i, \mathbf{u}_j)$ of a beam as a function of the generalized displacements. We write it for a beam parallel to the Ox_1 axis (see Fig. 4.5). The generalized displacements of a beam consist of the axial displacements (u_1, u_7), the normal deflections (u_2, u_3, u_8, u_9), and the angles of rotation ($u_4, u_5, u_6, u_{10}, u_{11}, u_{12}$) of the beam's ends (see Fig. 4.5). In these notations,

$$\mathbf{u}_i = (u_1, u_2, u_3, u_4, u_5, u_6), \mathbf{u}_j = (u_7, u_8, u_9, u_{10}, u_{11}, u_{12}),$$

and the elastic energy is computed in accordance with the formula,

$$E(\mathbf{u}_i, \mathbf{u}_j) = \frac{1}{2}\int_0^L D\left(\frac{u_1}{L} - \frac{u_2}{L}\right)^2 dy_1 \tag{4.3.39}$$

$$+\frac{1}{2}\int_0^L D_2\left[\left(\frac{u_2}{L^3} - \frac{u_8}{L^3}\right)(12l - 6L) + \left(\frac{u_4}{L^2} - \frac{u_{10}}{L^2}\right)(6l - 2L)^2\right]^2 dy_1$$

$$+\frac{1}{2}\int_0^L D_3\left[\left(\frac{u_3}{L^3}-\frac{u_9}{L^3}\right)(12l-6L)+\left(\frac{u_5}{L^2}-\frac{u_{11}}{L^2}\right)(6l-2L)^2\right]^2 dy_1$$

$$+\frac{1}{2}\int_0^L I\left(\frac{u_6}{L}-\frac{u_{12}}{L}\right)^2 dy_1 .$$

Here, D is the axial stiffness of the beam, D_2 and D_3 are the bending stiffnesses in the Oy_1y_2 and Oy_1y_3 planes, respectively, I is the torsional stiffness, and L denotes the length of the beam.

Formula (4.3.39) can be written in an arbitrary coordinate system; see Haug et al. (1986) for details.

Rod cellular structures

Consider a cellular structure made of rods. A rod works only in the tension-compression mode. Then $\mathbf{Q}$=0, M_α=0. The axial force in a rod is computed as follows:

$$\mathbf{N}_{ij}=\frac{E_{ij}(\mathbf{u}_i-\mathbf{u}_j,\mathbf{e}_{ij})\mathbf{e}_{ij}}{L_{ij}}, \tag{4.3.40}$$

where subscript ij refers to rod with the ends at the i th and j th nodes, E_{ij} is the rigidity of this rod, $\mathbf{e}_{ij}$ is its directing vector, L_{ij} means the length of the rod, and (,) means the scalar product.

Allowing for (4.3.40), (4.3.30) takes the form,

$$\sum_{j\in K_+}\frac{E_{ij}(\mathbf{u}_i-\mathbf{u}_j,\mathbf{e}_{ij})\mathbf{e}_{ij}}{L_{ij}}=0 . \tag{4.3.41}$$

Equation (4.3.41) is a system of linear algebraic equations with respect to the generalized displacements . It can be written in the form,

$$\mathbf{T}_r\{\mathbf{u}_i\}=\mathbf{b}^{\nu\alpha\beta} . \tag{4.3.42}$$

Matrix $\mathbf{T}_r$ is a fragment of matrix $\mathbf{T}_b$.

For a rod, (4.3.39) becomes

$$E(\mathbf{u}_i, \mathbf{u}_j) = \frac{1}{2}\int_0^L D\left(\frac{u_1}{L} - \frac{u_2}{L}\right)^2 dy_1 . \qquad (4.3.43)$$

4.4 The Homogenization Method Modified for Lattice Beams

In this section, a method for solving cellular problems and computing homogenized stiffnesses for lattice beams (see Fig.4.6) is presented. The method is similar to the method presented in Sect. 4.3. The main stage of the method will be constructing analogs of functions $\xi^{\nu\alpha\beta}(\mathbf{y})$.

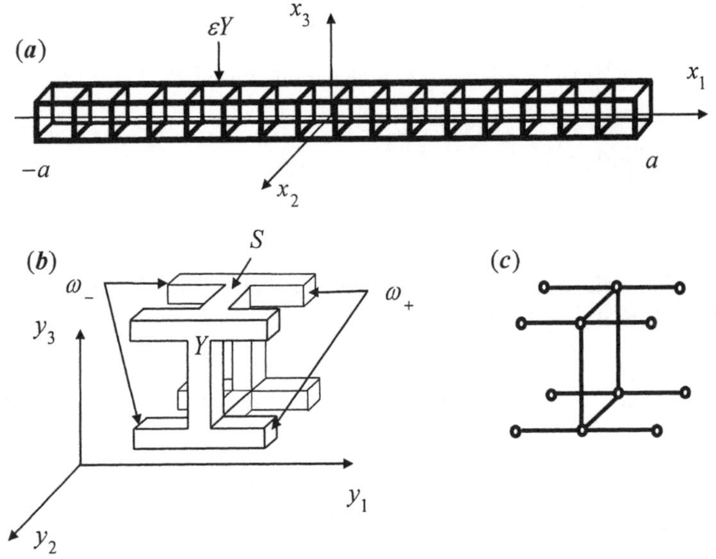

Fig. 4.6. (*a*) Beam-like framework; (*b*), its periodicity cell; and (*c*) and finite-elements model of the periodicity cell

Transformation of the cellular boundary value problem into a finite-dimensional problem for lattice beam-like structures

Cellular problems for a beam are the following (see Sect. 4.2):

$$\begin{cases} [c_{ijmn}(\mathbf{y})X^{\nu\alpha}_{m,n} + c_{ij11}(\mathbf{y})y^{\nu}_{\alpha}]_{,jy} = 0 \quad \text{in } Y, \\ [c_{ijmn}(\mathbf{y})X^{\nu\alpha}_{m,n} + c_{ij11}(\mathbf{y})y^{\nu}_{\alpha}]n_j = 0 \quad \text{on } S, \\ \mathbf{X}^{\nu\alpha}(\mathbf{y}) \text{ is periodic in } y_1 \text{ with periodicity cell } T, \\ \langle \mathbf{X}^{\nu\alpha} \rangle = 0; \end{cases} \tag{4.4.1}$$

$$\begin{cases} [c_{ijmn}(\mathbf{y})X^{3}_{m,n} + c_{ij\gamma1}(\mathbf{y})s_{\gamma}y_{\Gamma}]_{,jy} = 0 \quad \text{in } Y, \\ [c_{ijmn}(\mathbf{y})X^{3}_{m,n} + c_{ij\gamma1}(\mathbf{y})s_{\gamma}y_{\Gamma}]n_j = 0 \quad \text{on } S, \\ \mathbf{X}^{3}(\mathbf{y}) \text{ is periodic in } y_1 \text{ with periodicity cell } T, \\ \langle \mathbf{X}^{3} \rangle = 0. \end{cases} \tag{4.4.2}$$

After the cellular problem is solved, the homogenized characteristics are computed in accordance with the formula,

$$A^2_{\alpha\alpha} = \langle c_{ijkl}(\mathbf{y})[X^{\alpha}_{k,l} + y_{\alpha}\delta_{k1}\delta_{l1}][X^{\alpha}_{i,j} + y_{\alpha}\delta_{i1}\delta_{j1}]\rangle, \tag{4.4.3}$$

$$B = \langle c_{ijkl}(\mathbf{y})[X_{k,l} + y_{\Gamma}s_{\gamma}\delta_{k\gamma}\delta_{l1}][X_{i,j} + y_{\Gamma}s_{\gamma}\delta_{i\gamma}\delta_{j1}]\rangle. \tag{4.4.4}$$

ν, μ take values 0, 1; $\alpha, \beta, \gamma, \delta = 1$, 2; $\mathrm{B} = 2$ if $\beta = 3$; and $\mathrm{B} = 3$ if $\beta = 2$. Here, A^0_{11} is the axial stiffness, $A^1_{\alpha\beta}$ are the coupling stiffnesses, $A^2_{\alpha\beta}$ are the bending stiffnesses, and B means the torsional stiffness of the homogenized beam.

Transformation of cellular problems into problems of the theory of strength

Problems (4.4.1), (4.4.2) represent elastic problems of a special form. The specifics of the problems are related to the free terms and boundary conditions. The boundary conditions do not raise problem because they can be easily reformulated in terms of the theory of finite-dimensional structures.

Then, the specifics of cellular problems are related to the terms, $[c_{ij11}(\mathbf{y})y_\alpha^\nu]_{,j}$ ($\nu = 0,\ 1$) and $[c_{ij\gamma 1}(\mathbf{y})s_\gamma y_\Gamma]_{,j}$. We demonstrate that they can be eliminated from cellular problems (4.4.1) and (4.4.2). We can eliminate the term $[c_{ij11}(\mathbf{y})]_{,j}$ from problem (4.4.1) by a known change of function. We introduce

$$\mathbf{W}(\mathbf{y}) = \mathbf{X}^0(\mathbf{y}) + y_1\mathbf{e}_1 \ . \tag{4.4.5}$$

After that, (4.4.1), (4.4.2) with respect to function $\mathbf{W}(\mathbf{y})$ take the form of the usual equations of elasticity theory with no mass forces.

Let us demonstrate that a similar procedure can be used for $\nu = 1$ (for problems of bending and torsion). The problem is equivalent to the problem of the existence of displacements $\xi^{1\alpha}(\mathbf{y})$ and $\xi^\gamma(\mathbf{y})$ such that the following conditions are satisfied:

$$c_{ijmn}(\mathbf{y})e_{mn}^{1\alpha} + c_{ij11}(\mathbf{y})y_\alpha = 0 \quad \text{for problem (4.4.1);} \tag{4.4.6}$$

$$c_{ijmn}(\mathbf{y})e_{mn}^{\gamma} + s_\gamma y_\Gamma c_{ij\gamma 1}(\mathbf{y}) = 0 \quad \text{for problem (4.4.2);} \tag{4.4.7}$$

for strains $e_{kl} = (1/2)(\partial\xi_k / \partial x_l + \partial\xi_l / \partial x_k)$ corresponding to $\xi^{1\alpha}(\mathbf{y})$ and $\xi^\gamma(\mathbf{y})$, respectively.

Solving (4.4.6) and (4.4.7) with respect to $\{e_{mn}\}$, we obtain

$$e_{mn}^{1\alpha} = -c_{mnij}^{-1}(\mathbf{y})c_{ij11}(\mathbf{y})y_\alpha , \tag{4.4.8}$$

$$e_{mn}^{\gamma} = -c_{mnij}^{-1}(\mathbf{y})c_{ij\gamma 1}(\mathbf{y})s_\gamma y_\Gamma , \tag{4.4.9}$$

where c_{klij}^{-1} means the tensor inverse with respect to tensor c_{klij}. It follows from (4.4.8) and (4.4.9) that

$$e_{kl}^{1\alpha} = -\delta_{k1}\delta_{l1}y_\alpha , \tag{4.4.10}$$

$$e^{\gamma}_{kl} = -\delta_{k\gamma}\delta_{l1}s_{\gamma}y_{\Gamma} \ . \tag{4.4.11}$$

As seen, the problem is reduced to the problem of compatibility of strains determined by (4.4.10) and (4.4.11). The equations of compatibility (4.3.7) are satisfied for strains (4.4.10) and (4.4.11), all of which represent linear functions. Thus, displacements $\xi^{1\alpha}(\mathbf{y})$ satisfying (4.4.10) and $\xi^{\gamma}(\mathbf{y})$ satisfying (4.4.11) exist.

Constructing of the functions $\xi^{\alpha}(\mathbf{y})$ *and* $\xi^{\gamma}(\mathbf{y})$

We obtain explicit formulas for functions $\xi^{1\alpha}(\mathbf{y})$ and $\xi^{\gamma}(\mathbf{y})$.

Function $\xi^{13}(\mathbf{y})$ ($\alpha = 3$)

In this case,

$$e_{11} = -y_3 \text{, and } e_{ij} = 0 \quad \text{for } ij \neq 11 . \tag{4.4.12}$$

Comparing (4.4.12) and (4.3.8), we find that

$$\xi^{13}(\mathbf{y}) = -\xi^{122}(\mathbf{y}) .$$

Thus,

$$\xi^{13}_1(\mathbf{y}) = -y_1 y_3 , \tag{4.4.13}$$

$$\xi^{13}_2(\mathbf{y}) = \frac{y_1^2}{2} ,$$

$$\xi^{13}_3(\mathbf{y}) = 0 .$$

Function $\xi^{12}(\mathbf{y})$ ($\alpha = 2$)

Function $\xi^{12}(\mathbf{y})$ can be obtained from function $\xi^{13}(\mathbf{y})$ by replacing index $\alpha = 2$ with α =3. Then,

$$\xi^{12}_1(\mathbf{y}) = -y_1 y_2 , \ \xi^{12}_3(\mathbf{y}) = \frac{-y_1^2}{2} , \ \xi^{12}_2(\mathbf{y}) = 0 . \tag{4.4.14}$$

Function $\xi^\gamma(\mathbf{y})$

The strains corresponding to function $\xi^\gamma(\mathbf{y})$ are the following:

$$e_{12} = e_{21} = -y_3 + y_2, \; e_{13} = e_{31} = y_2, \tag{4.4.15}$$

$$\text{and } e_{ij} = 0 \text{ for } ij \neq 12, 21, 13, 31.$$

The solution of problem (4.4.15) is the following:

$$\xi_1^\gamma(\mathbf{y}) = 0, \; \xi_2^\gamma(\mathbf{y}) = 2y_1 y_3, \; \xi_3^\gamma(\mathbf{y}) = -2y_1 y_2. \tag{4.4.16}$$

Denote by $\xi(\mathbf{y})$ one of the following functions:

$$\xi^{1\alpha}(\mathbf{y}), \; \xi^\gamma(\mathbf{y}), \; \xi^0(\mathbf{y}) = y_1 \mathbf{e}_1, \tag{4.4.17}$$

where displacements $\xi^{1\alpha}(\mathbf{y})$ are determined by (4.4.13) for $\alpha = 3$, by (4.4.14) for $\alpha = 2$, and $\xi^\gamma(\mathbf{y})$ are determined by (4.4.16). Function $\xi^0(\mathbf{y})$ corresponds to cellular problem (4.4.1) with $\nu = 0$. Using the notations introduced, we change the functions in accordance with the formula,

$$\mathbf{W}(\mathbf{y}) = \mathbf{X}^0(\mathbf{y}) + \xi^0(\mathbf{y}) \quad \text{for (4.4.1) } \nu = 0;$$

$$\mathbf{W}(\mathbf{y}) = \mathbf{X}^{1\alpha}(\mathbf{y}) + \xi^{1\alpha}(\mathbf{y}) \quad \text{for (4.4.1) } \nu = 1;$$

$$\mathbf{W}(\mathbf{y}) = \mathbf{X}^3(\mathbf{y}) + \xi^\gamma(\mathbf{y}) \quad \text{for (4.4.2)},$$

and can write the cellular problems in the form

$$\begin{cases} [c_{ijmn}(\mathbf{y}) W_{m,n}]_{,j} = 0 & \text{in } Y, \\ [c_{ijmn}(\mathbf{y}) W_{m,n}] n_j = 0 & \text{on } S, \\ \mathbf{W}(\mathbf{y}) \text{ is periodic in } y_1 \text{ with periodicity cell } T, \\ \langle \mathbf{W} - \xi(\mathbf{y}) \rangle = 0. \end{cases} \tag{4.4.18}$$

Method for solving cellular problems and computing of homogenized stiffnesses

Cellular problem (4.4.18) is similar to problem (4.3.28). It can be solved in the way described in Sect. 4.3 [see subsection "Method for solving cellular problem (4.3.28)"]. The homogenized characteristics can be computed using (4.4.3), (4.4.4) with suitable functions $\xi^{1\nu\alpha}(\mathbf{y})$, $\xi^{\gamma}(\mathbf{y})$, and $\xi^{0}(\mathbf{y})$.

Examples

We give two examples. The examples are simple, and they are given to illustrate the idea of the method presented above. We consider, for simplicity, planar structures.

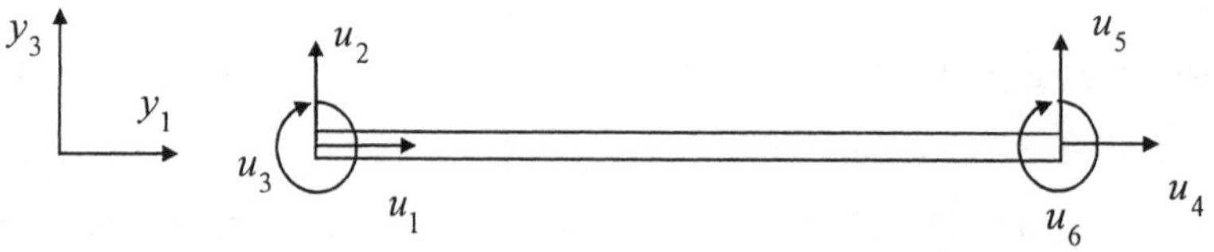

Fig. 4.7. The generalized displacements for a planar beam

The generalized displacements for a planar beam are (see Fig. 4.7) the following: the displacements of the ends of the beam u_1, u_2, u_4, u_5 and the rotational angles of the ends of the beam u_3, u_6.

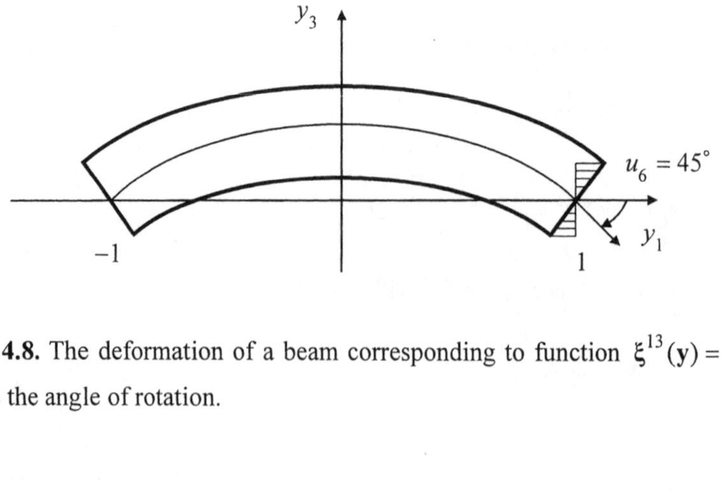

Fig. 4.8. The deformation of a beam corresponding to function $\xi^{13}(\mathbf{y}) = y_1 y_3 \mathbf{e}_1$; u_6 is the angle of rotation.

The rotations corresponding to function $\xi^{13}(\mathbf{y}) = y_1 y_3 \mathbf{e}_1$ are equal to 45° (Fig. 4.8). This is not a small value. It does not create any mathematical problems. Nevertheless, this value is not usual from the mechanical point of view. We introduce function $\xi^{\lambda}(\mathbf{y}) = \lambda y_1 y_3 \mathbf{e}_1$, where λ is a parameter, which will be assumed to be a small number, and investigate the cellular problems with function $\xi^{\lambda}(\mathbf{y})$ instead of $\xi(\mathbf{y})$. The solution of this new cellular problem is

$$\mathbf{W}^{\lambda}(\mathbf{y}) = \lambda \mathbf{X}(\mathbf{y}) + \xi^{\lambda}(\mathbf{y}) = \lambda \mathbf{W}^{\lambda}(\mathbf{y}),$$

and the new cellular problem is a problem of elasticity theory for small deformations.

The elastic energy for the new cellular problem is

$$E^{\lambda} = E\lambda^2 .$$

The coefficient E multiplied by two and then divided by the measure of the periodicity cell is equal to the value of the corresponding homogenized stiffness [see formula (4.3.38)].

One beam

Consider the periodicity cell formed by one beam with ends at points $y_1 = -1$ and $y_1 = 1$. We computer the bending stiffness of the beam using the method developed in the book. Evidently, any correct method must give the result predicted by classical theory. The generalized displacements have the following meanings:

$$u_1 = u(-1), \quad u_4 = u(1),$$

$$u_2 = w(-1), \quad u_5 = w(1),$$

$$u_3 = w'(-1), \quad u_6 = w'(1),$$

where prime means the derivative with respect to variable y_1, u means the axial displacement, and w means the normal deflection.

The periodicity conditions for a beam take the form,

$$[u_1 - u_3] + [\xi_1^{\lambda}] = 0,$$

$$[u_2 - u_4] + [\xi_2^{\lambda}] = 0,$$

$$[u_3 - u_6] + [\frac{\partial}{\partial y_3} \xi_1^\lambda] = 0,$$

([] means "jump", $[f] = f(1) - f(-1)$). This system is system (4.3.33) written for the partial case - one planar beam.

Taking into account that

$$\xi_1^\lambda (y_1 = \pm 1, y_3 = 0) = \lambda y_1 y_3 \mathbf{e}_1 \Big|_{y_3=0} = 0,$$

$$\xi_3^\lambda (y_1 = \pm 1, y_3 = 0) = 0,$$

$$\frac{\partial}{\partial y_3} \xi_1^\lambda (y_1 = \pm 1, y_3 = 0) = \lambda y_1 \Big|_{y_1 = \pm 1} = \pm \lambda,$$

we arrive at the problem

$$u' = 0, u(-1) = u(1) = 0,$$

$$w'''' = 0, w(-1) = w(1) = 0, w'(1) - w'(-1) = 2\lambda.$$

The solution of this problem is

$$u = 0, \; w = \frac{\lambda}{2}(y_1^2 - 1).$$

The energy of the beam, multiplied by two and then divided by its length (which is equal to 2), is

$$E^\lambda = \frac{1}{2} \int_{-1}^{1} D\lambda^2 dy_1 = D\lambda^2,$$

where D is the beam bending stiffness computed in accordance with the classical formula. Thus, the stiffness of a uniform beam computed with the developed method and the stiffness computed with the classical theory are the same. This means that the method developed is in agreement with the classical theory of beams.

Lattice beam with I-shaped periodicity cell

Consider a lattice beam that periodicity cell is shown in Fig. 4.9.

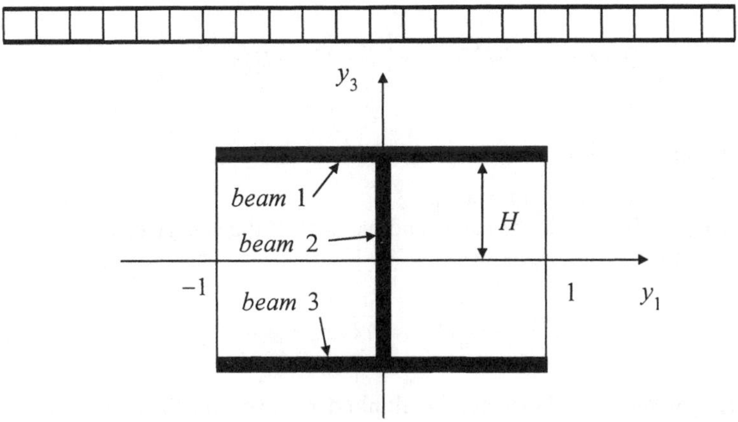

Fig. 4.9. Lattice beam with I-shaped periodicity cell

For this structure,

$$\xi_1^\lambda(y_1 = \pm 1, y_3 = H) = \pm H,$$

$$\xi_3^\lambda(y_1 = \pm 1, y_3 = H) = 0,$$

$$\frac{\partial}{\partial y_3}\xi_1^\lambda(y_1 = \pm 1, y_3 = H) = \pm\lambda .$$

For beam 1, we obtain the problem,

$$u' = 0,\ u(-1) = -\lambda H,\ u(1) = \lambda H,$$

$$w'''' = 0,\ w(-1) = w(1) = 0,\ w'(1) - w'(-1) = 2\lambda .$$

The solution of this problem is

$$u = \lambda H y_1,\ w = \frac{\lambda}{2}(y_1^2 - 1).$$

The energy of beam 1 is

$$\int_{-1}^{1} [D_0(\lambda H)^2 + D\lambda^2]dy_1 = \frac{1}{2}(D_0 H^2 + D)\lambda^2 ,$$

where D_0 and D are axial (tension) and bending stiffnesses computed in accordance with the classical formula.

The energy of beam 3 is the same, and we neglect the energy of beam 2. Then, the energy of the periodicity cell is

$$E^\lambda = (D_0 H^2 + D)\lambda^2 .$$

The energy of the periodicity cell, multiplied by two and then divided by the measure of the cell (in the case considered, it is equal to $4H$; see Fig. 4.9), is equal to the bending stiffness of an I-shaped lattice, and it is

$$2\left(D_0 H + \frac{D}{H}\right). \tag{4.4.19}$$

In the computations above, we neglected the thickness of beams compared with the dimension of the periodicity cell. This means that (4.4.19) cannot be used for a small distance H between beams. Neglecting D/H, we obtain the well-known formula for the bending stiffness of a beam with a double-T cross section.

4.5 Review of Software Suitable for Homogenization Procedures

All the homogenization procedures considered above consist of two stages: solution of a cellular problem and computation of homogenized characteristics.

As follows from Sect. 4.4 and 4.5, a cellular problem represents a special type of elasticity problem with periodicity boundary conditions with respect to some variables. It can be written in the general form,

$$\begin{cases} [c_{ijmn}(\mathbf{y})W_{m,n}]_{,j} = 0 \quad \text{in } Y, \\ c_{ijmn}(\mathbf{y})W_{m,n}n_j = 0 \quad \text{on } S, \\ \langle \mathbf{W} \rangle = 0, \end{cases} \tag{4.5.1}$$

with periodicity conditions

$$\mathbf{W}(\mathbf{y}) - \xi(\mathbf{y}) \text{ periodic in } \begin{cases} \mathbf{y} & \text{for a 3-D composite,} \\ y_1, y_2 & \text{for a 2-D composite,} \\ y_1 & \text{for a 1-D composite} \end{cases} \tag{4.5.2}$$

with periodicity cell T.

There exist the following cellular problems in the theory of homogenization of elastic structures (the problems are described in detail in Sect. 1.2, 2.8 and 3.8).

One cellular problem for 3-D (solid-like) composite structures.

Two cellular problems for 2-D (plate-like) composite structures:

- the cellular problem, corresponding to in-plane strain,
- the cellular problem, corresponding to bending.

Three cellular problems for 1-D (beam-like) composite structures:

- the cellular problem, corresponding to axial strain,
- the cellular problem, corresponding to bending,
- the cellular problem, corresponding to torsion.

All the problems mentioned above can be written in the form (4.5.1) with a suitable function $\xi(\mathbf{y})$. Function $\xi(\mathbf{y})$ for every type of cellular problem was described in Sects. 4.3 and 4.4.

After the cellular problem (4.5.1), (4.5.2) is solved, we compute the homogenized characteristics. All formulas for computing the homogenized characteristics can be written in the following form:

$$(mesT)^{-1} \int_Y g(W_{i,j}(\mathbf{y}), \xi(\mathbf{y}))d\mathbf{y} . \tag{4.5.3}$$

For every type of homogenized characteristics, g is a known function (it can be found in Sects. 1.2, 2.2, and 3.2).

In addition, the local strains and stresses in the structure can be computed. Formulas for computing local strains and stresses have the following general form:

$$e_{ij}^{\varepsilon} = T_{ijmn}(\mathbf{x}/\varepsilon)u_{m,nx}^{(0)}(\mathbf{x}), \tag{4.5.4}$$

$$\sigma_{ij}^{\varepsilon} = S_{ijmn}(\mathbf{x}/\varepsilon)u_{m,nx}^{(0)}(\mathbf{x}),$$

where $\mathbf{u}^{(0)}$ are the homogenized displacements. Tensors $T_{ijmn}(\mathbf{x}/\varepsilon)$ and $S_{ijmn}(\mathbf{x}/\varepsilon) = c_{ijkl}(\mathbf{x}/\varepsilon)T_{klmn}(\mathbf{x}/\varepsilon)$ are called the tensors of local strains and local stresses, respectively. They also are called the tensors of concentration of local strains and local stresses in composites.

We use the ideas of functional analysis (see, e.g., Kolmogorov and Fomin, 1970; Ladigenskaya, 1973), and can say that the differential operator in all cellular problems is the same and solutions of cellular problems can be obtained in the form

$$\mathbf{W} = F(\xi), \tag{4.5.5}$$

where F is an operator inverse with respect to the differential operator from (4.5.1). The operator F is the same for all types of cellular problems under consideration. This note is not useful to engineers directly. But it is a fruitful note for numerical analysis of the problem. Operator F can be constructed numerically, of course, not in the form of a usual function, but in the form of a numerical procedure; see Bathe (1982), Samarskii (1977), and Lions (1978). The numerical procedure (the corresponding computer subprogram) will be the same for all types of cellular problems.

Solution of cellular problems using FEM

We will discuss briefly the numerical solution of a cellular problem with commercial software. Problem (4.5.1) can be solved by a finite elements method using software, which allows formulating the periodicity boundary conditions. Many of the modern commercial finite element programs have the option "Cyclic symmetry analysis" (or similar), which allows solving periodicity problems.

The stages of the numerical solution of the problem are the following:

The preprocessor stages

1. Describe the type and geometry of periodicity cell Y.
2. Describe the type of the finite elements and generate the finite element mesh.
3. Input the material characteristics $a_{ijkl}(\mathbf{y})$.
4. Input the boundary conditions on the free surface S ($S = \varnothing$ for 3-D composite).
5. Input the condition $\langle \mathbf{W} \rangle = 0$.

6. Input the periodicity conditions (4.5.2) with a suitable function $\xi(\mathbf{y})$.
7. Send the problem to the FEM processor.

The postprocessor stages

8. Compute homogenized characteristics (4.5.3).
9. Compute tensors of local strains and stresses (4.5.4).

Review of FEM software suitable for solution of cellular problems

MSC/NASTRAN software

In NASTRAN (Introduction to MSC/NASTRAN, 1994) a periodicity conditions can be formulated using the option "Cyclic symmetry analysis." This option can be used to formulate the boundary conditions (constraint equations) of the form,

$$\sum_{a=1}^{A} A_a \mathbf{X}_a = 0,$$

where $\{\mathbf{X}_a\}$ are the generalized displacements of a node.

The periodicity conditions can be written in the form $\mathbf{X}_{a+} - \mathbf{X}_{a-} = 0$ or 1. It means that the coefficient $A_{a+} = 1$, $A_{a-} = -1$.

The NASTRAN library includes solid finite elements and beam and rod finite elements. Beam and rod finite elements can be used to carry out the homogenization procedure for lattice plates and beams described in Sect. 4.3 and 4.2.

ABAQUS software

ABAQUS (www.abaqus.com) software has the option "Cyclic symmetry analysis," which can be used to solve cellular problems.

ANSYS software

ANSIS software (*ANSYS*, 1992) has the option "Constraint equations ", which can be used to solve periodic problems in a way similar to that described in the paragraph devoted to NASTRAN software.

The insufficiencies of all the programs for homogenization are the same:

- One has to input the periodicity conditions by hand.
- There are no tools for computing homogenized characteristics (the second stage of the homogenization procedure).

These problems can be solved by writing pre- and postprocessor programs.

Solution of finite-dimensional cellular problems

Finite-dimensional cellular problems [(4.3.33) for plates made of beams, (4.3.42) for plates made of rods, and similar problems for lattice beams] have the form,

$$\mathbf{T}\{\mathbf{u}_i\} = \mathbf{b}(\xi), \tag{4.5.6}$$

where the matrix $\mathbf{T}$ ($\mathbf{T} = \mathbf{T}_b$ or $\mathbf{T} = \mathbf{T}_r$) does not depend on the type of the cellular problem. The solution of problem (4.5.6) is

$$\{\mathbf{u}_i\} = \mathbf{T}^{-1}\mathbf{b}(\xi), \tag{4.5.7}$$

where matrix $\mathbf{T}^{-1}$ is inverse with respect to matrix $\mathbf{T}$. In the case considered, operator $F = \mathbf{T}^{-1}\mathbf{b}$ is a matrix. The inverse matrix $\mathbf{T}^{-1}$ for a problem of small dimension can be computed in explicit form. Usually, the dimension of the matrix $\mathbf{T}$ is some tens, and computer programs of linear algebra must be used to compute matrix $\mathbf{T}^{-1}$.

Computation of homogenized stiffnesses using the exact solution of cellular problems

In the book by Kalamkarov and Kolpakov (1997), exact formulas for computing of effective stiffnesses of plates of complex structure were presented. The following types of plates were considered:

laminated plates;
ribbed plates;
wafer plates;
honeycomb-like plates (sandwich plates with honeycomb filler);
network plates;
fiber-reinforced plates.

The computational formulas have a complex form for some of the structures mentioned. The effective stiffnesses for all the above-mentioned types of structures can be computed using DESIGNER software described in the book by Kalamkarov and Kolpakov (1997). The software, DESIGNER, can be downloaded from http://mechv.me.turns.ca/akalam/DESIGNER.

Appendix A. Stiffnesses of Plates and Beams in Different Coordinate Systems

The quantities (moments and stiffnesses), introduced in Chaps. 2 and 3, and forming the models of plates and beams, depend on the taking of a coordinate system. The same dependencies take place in the classical theory developed for uniform plates and beams made of homogeneous materials (see, e.g., Appendix G in the book by Washizy, 1982). But, in classical theories, the problem of choice of coordinate system is not very actual because there exist the preferable (related to neutral plane or neutral axis) coordinate systems, which are normally used. For plates and beams of complex structure, neutral planes or axes can correspond no mechanical objects. Nevertheless, for plates and beams of complex structure, the problem of choice of coordinate system can be solved in agreement with the idea about the independence of mechanical behavior of structure on the choice of coordinate system.

In this section it is demonstrated that:

1. The dependence of the stiffnesses on the coordinate system is a natural feature of the stiffnesses.
2. There exist invariant forms of the homogenized models of plates and beams.
3. There exist simplest forms of the homogenized models.

A1 Homogenized Stiffnesses of a Plate in Different Coordinate Systems

Consider cellular problem (2.2.42) for plate in two coordinates systems ("old" and "new") related among themselves by the following transformation:

$$y_\alpha \to y_\alpha \ (\alpha{=}1,2), \ \ y_3 \to y_3 + h. \tag{A1}$$

The cellular problem (2.2.42) in the "new" coordinate system takes the form

$$\begin{cases} [c_{ijmn}(\mathbf{y})X^{1\alpha\beta}_{m,ny} - (y_3+h)c_{ij\beta\alpha}(\mathbf{y})]_{,j} = 0 \ \text{in } Y, \\ [c_{ijmn}(\mathbf{y})X^{1\alpha\beta}_{m,ny} - (y_3+h)c_{ij\beta\alpha}(\mathbf{y})]n_j = 0 \ \text{on } S, \\ \mathbf{X}^{1\alpha\beta}(\mathbf{y}) \ \text{is periodic in } y_1, y_2 \text{ with periodicity cell } T. \end{cases} \tag{A2}$$

Solution $\mathbf{X}^{1\alpha\alpha}(h,\mathbf{y})$ of problem (A2) is related to solutions $\mathbf{X}^{0\alpha\alpha}(\mathbf{y})$ and $\mathbf{X}^{1\alpha\alpha}(\mathbf{y})$ of cellular problems (2.2.31), (2.2.42) in the "old" coordinate system as follows:

$$\mathbf{X}^{1\alpha\alpha}(h,\mathbf{y}) = \mathbf{X}^{1\alpha\alpha}(\mathbf{y}) + h\mathbf{X}^{0\alpha\alpha}(\mathbf{y}). \tag{A3}$$

Argument h indicates that the function is calculated in the "new" coordinate system. Functions without argument h are computed in the "old" coordinate system.

Using (2.2.47) for computation of the homogenized stiffnesses and formula (A3), we obtain the following formulas relating the stiffnesses $A^2_{\alpha\alpha\alpha\alpha}$ in the "old" and "new" coordinate systems:

$$A^2_{\alpha\alpha\alpha\alpha}(h) = \langle [c_{\alpha\alpha\alpha\alpha}(\mathbf{y})(y_3 + h)$$

$$- c_{\alpha\alpha mn}(\mathbf{y})(X^{1\alpha\alpha}_{m,ny}(\mathbf{y}) + hX^{0\alpha\alpha}_{m,ny}(\mathbf{y})](y_3 + h)\rangle .$$

After algebraic transformations, we obtain,

$$A^2_{\alpha\alpha\alpha\alpha}(h) = A^2_{\alpha\alpha\alpha\alpha}(h) + 2hA^1_{\alpha\alpha\alpha\alpha} + h^2 A^0_{\alpha\alpha\alpha\alpha}. \tag{A4}$$

For coupling and in-plane stiffnesses $A^1_{\alpha\alpha\alpha\alpha}(h)$ and $A^0_{\alpha\alpha\alpha\alpha}(h)$, the following formulas can be derived in a similar way:

$$A^1_{\alpha\alpha\alpha\alpha}(h) = A^1_{\alpha\alpha\alpha\alpha} + hA^0_{\alpha\alpha\alpha\alpha},\quad A^0_{\alpha\alpha\alpha\alpha}(h) = A^0_{\alpha\alpha\alpha\alpha}. \tag{A5}$$

Invariant form of homogenized model of a plate

Stiffnesses of a plate, determined as above, depend on the choice of coordinate system. Note, that moments, introduced by the formula,

$$M_{\alpha\beta} = \langle \sigma^{\varepsilon}_{\alpha\beta} y_3 \rangle \tag{A6}$$

($\sigma^{\varepsilon}_{\alpha\beta}$ is the tensor of local stresses in plate), also depend on the choice of coordinate system.

In order to derive the invariant model of plate, we modify the definition of moments. Using in-plate stresses

$$N_{\alpha\beta} = \langle \sigma^{\varepsilon}_{\alpha\beta} \rangle , \tag{A6}$$

we introduce the following quantities:

$$M^0_{\alpha\beta} = M_{\alpha\beta} - \tau\langle y_3\rangle N_{\alpha\beta}, \text{ where } \tau = \frac{mesY}{mesS}, \tag{A7}$$

which will be called normalized moments to distinguish them from the normally used moments $M_{\alpha\beta}$. The number τ in (A7) is taken in such way, that the equality $\tau\langle y_3 + h\rangle = \tau\langle y_3\rangle + h$ be satisfied.

The normalized moments are invariants (do not depend on change (A1) of coordinate system). Really, in the "new" coordinate system we have,

$$M^0_{\alpha\beta}(h) = M_{\alpha\beta}(h) - \tau\langle y_3 + h\rangle N_{\alpha\beta}$$

$$= M_{\alpha\beta} + hN_{\alpha\beta} - \tau\langle y_3\rangle N_{\alpha\beta} - \tau\langle h\rangle N_{\alpha\beta}$$

$$= M_{\alpha\beta} - \tau\langle y_3\rangle N_{\alpha\beta} + h(1 - \tau\frac{mesS}{mesY})$$

$$= M_{\alpha\beta} - \tau\langle y_3\rangle N_{\alpha\beta} = M^0_{\alpha\beta},$$

where $M^0_{\alpha\beta}$ is a moment computed in the "old" coordinate system.

The constitutive homogenized equations (2.2.45), (2.2.46) can be written in the form

$$N_{\alpha\beta}(h) = A^0_{\alpha\beta\gamma\delta}(h)e_{\gamma\delta} + A^1_{\alpha\beta\gamma\delta}(h)\rho_{\gamma\delta},$$

$$M^0_{\alpha\beta}(h) = A^1_{\alpha\beta\gamma\delta}(h)e_{\gamma\delta} + A^2_{\alpha\beta\gamma\delta}(h)\rho_{\gamma\delta}$$

$$- \tau\langle y_3\rangle(A^0_{\alpha\beta\gamma\delta}(h)e_{\gamma\delta} + A^1_{\alpha\beta\gamma\delta}(h)\rho_{\gamma\delta}).$$

These equations can be rewritten as

$$N_{\alpha\beta}(h) = A^0_{\alpha\beta\gamma\delta}(e_{\gamma\delta} + h\rho_{\gamma\delta}) + A^1_{\alpha\beta\gamma\delta}(h)\rho_{\gamma\delta};$$

$$M^0_{\alpha\beta}(h) = A^1_{\alpha\beta\gamma\delta}e_{\gamma\delta} + hA^1_{\alpha\beta\gamma\delta}e_{\gamma\delta}$$

$$+ A^2_{\alpha\beta\gamma\delta}\rho_{\gamma\delta} + 2hA^2_{\alpha\beta\gamma\delta}\rho_{\gamma\delta} + h^2 A^0_{\alpha\beta\gamma\delta}\rho_{\gamma\delta}$$

$$- \tau\langle y_3 + h\rangle(A^0_{\alpha\beta\gamma\delta}e_{\gamma\delta} + A^1_{\alpha\beta\gamma\delta}\rho_{\gamma\delta} + hA^0_{\alpha\beta\gamma\delta}\rho_{\gamma\delta})$$

$$= A^1_{\alpha\beta\gamma\delta}(e_{\gamma\delta} + h\rho_{\gamma\delta}) + A^2_{\alpha\beta\gamma\delta}\rho_{\gamma\delta} - \tau\langle y_3\rangle(A^0_{\alpha\beta\gamma\delta}(e_{\gamma\delta} + h\rho_{\gamma\delta}) + A^1_{\alpha\beta\gamma\delta}\rho_{\gamma\delta})$$

$$+ hA^0_{\alpha\beta\gamma\delta}e_{\gamma\delta} + hA^1_{\alpha\beta\gamma\delta}\rho_{\gamma\delta} + h^2 A^1_{\alpha\beta\gamma\delta}\rho_{\gamma\delta}$$

$$- \tau\langle y_3\rangle(A^0_{\alpha\beta\gamma\delta}e_{\gamma\delta} + hA^1_{\alpha\beta\gamma\delta}\rho_{\gamma\delta} + h^2 A^1_{\alpha\beta\gamma\delta}\rho_{\gamma\delta}).$$

The sum

$$hA^0_{\alpha\beta\gamma\delta}e_{\gamma\delta} + hA^1_{\alpha\beta\gamma\delta}\rho_{\gamma\delta} + h^2 A^1_{\alpha\beta\gamma\delta}\rho_{\gamma\delta}$$

$$- \tau\langle y_3\rangle(A^0_{\alpha\beta\gamma\delta}e_{\gamma\delta} + hA^1_{\alpha\beta\gamma\delta}\rho_{\gamma\delta} + h^2 A^1_{\alpha\beta\gamma\delta}\rho_{\gamma\delta}) = 0.$$

As a result,

$$N_{\alpha\beta}(h) = A^0_{\alpha\beta\gamma\delta}(e_{\gamma\delta} + h\rho_{\gamma\delta}) + A^1_{\alpha\beta\gamma\delta}(h)\rho_{\gamma\delta},$$

$$M^0_{\alpha\beta}(h) = A^1_{\alpha\beta\gamma\delta}(e_{\gamma\delta} + h\rho_{\gamma\delta}) + A^2_{\alpha\beta\gamma\delta}\rho_{\gamma\delta}$$

$$- \tau\langle y_3\rangle[A^0_{\alpha\beta\gamma\delta}(e_{\gamma\delta} + h\rho_{\gamma\delta}) + A^1_{\alpha\beta\gamma\delta}\rho_{\gamma\delta}].$$

Equations (A8), considered with respect to the variables

$$N_{\alpha\beta},\ M^0_{\alpha\beta},\ e_{\gamma\delta} + h\rho_{\gamma\delta},\ \rho_{\gamma\delta}, \tag{A9}$$

are invariant with respect to change (A1) of coordinate system.

The homogenized equations of equilibrium (2.2.48)–(2.2.50) can be written in the form

$$N^{(-2)}_{\alpha\beta,\beta x} = \langle f_\alpha \rangle + \langle g_\alpha \rangle_S , \tag{A10}$$

$$M^{0}_{\alpha\beta,\alpha x\beta x} = (\langle f_\alpha y_3 \rangle + \langle g_\alpha y_3 \rangle_S)_{,\alpha_x} + \langle f_3 \rangle + \langle g_3 \rangle_S$$

$$-\tau \langle y_3 \rangle (\langle f_\alpha \rangle + \langle g_\alpha \rangle_S)_{,\alpha x} .$$

Equations (A10) are invariant with respect to change (A1) of the coordinate system.

Thus, we write all the equations, forming the homogenized model of plate, in a form invariant with respect to change (A1) of coordinate system.

The simplest form of homogenized constitutive equations of a plate

Let us analyze formulas (A4), (A5). It is seen from (A4), that the bending stiffnesses take the minimal value when

$$h = -\frac{A^1_{\alpha\alpha\alpha\alpha}}{A^0_{\alpha\alpha\alpha\alpha}} . \tag{A11}$$

It is seen from (A5), that h determined by (A11) gives, simultaneously, zero value to the coupling (out-of-plane) stiffness $A^1_{\alpha\alpha\alpha\alpha}$. The equation (A11) formally introduces the neutral plane $y_3 = h$ for a plate of complex structure (composite, lattice, net, and so on).

A2 Homogenized Stiffnesses of a Beam in Different Coordinate Systems

Consider cellular problem for beam (3.2.41) in two coordinates systems ("old" and "new") related among themselves by the following transformation:

$$y_\alpha \to y_\alpha + h_\alpha \ (\alpha=1,2), \ \ y_3 \to y_3 . \tag{A12}$$

Cellular problem (3.2.41) for $\nu = 1$ in the "new" coordinate system takes the form

$$\begin{cases} [c_{ijmn}(\mathbf{y})X^{1\alpha}_{m,ny} + c_{ij11}(\mathbf{y})(y_\alpha + h_\alpha)]_{,jy} = 0 \text{ in } Y, \\ [c_{ijmn}(\mathbf{y})X^{1\alpha}_{m,ny} + c_{ij11}(\mathbf{y})(y_\alpha + h_\alpha)]n_j = 0 \text{ on } S, \\ \mathbf{X}^{1\alpha}(\mathbf{y}) \text{ is periodic in } y_1 \text{ with period } T. \end{cases} \tag{A13}$$

Solution $\mathbf{X}^{1\alpha}(h,\mathbf{y})$ of problem (A13) is related to solutions $\mathbf{X}^{1\alpha}(\mathbf{y})$ and $\mathbf{X}^{01}(\mathbf{y})$ of cellular problems (3.2.29), (3.2.41) in the "old" coordinate system as follows:

$$\mathbf{X}^{1\alpha}(h,\mathbf{y}) = \mathbf{X}^{1\alpha}(\mathbf{y}) + h_\alpha \mathbf{X}^{01}(\mathbf{y}). \tag{A14}$$

From (3.2.48) and (A14), we obtain the formulas relating the stiffnesses in coordinate systems (A12),

$$A^2_{\alpha\alpha}(h) = \langle\{c_{\gamma\delta\alpha\beta}(\mathbf{y})(y_\alpha + h_\alpha) \tag{A15}$$

$$- c_{11mn}(\mathbf{y})[X^{1\alpha}_{m,ny}(\mathbf{y}) + hX^{01}_{m,ny}(\mathbf{y})]\}(y_\alpha + h_\alpha)\rangle.$$

After some transformations, we obtain the equality

$$A^2_{\alpha\alpha}(h) = A^2_{\alpha\alpha} + 2hA^1_\alpha + h^2 A. \tag{A15}$$

For coupling and axial stiffnesses $A^1_\alpha(h)$ and $A(h)$ the following equations can be derived:

$$A^1_\alpha(h) = A^1_\alpha + hA, \; A(h) = A. \tag{A16}$$

Invariant form of homogenized model of a beam

The stiffnesses of a beam, determined as above, depend on the choice of coordinate system. The moments, introduced by the formula

$$M_\alpha = \langle \sigma^\varepsilon_{11} y_\alpha \rangle \tag{A17}$$

($\sigma^\varepsilon_{\alpha\beta}$ is the tensor of local stresses in beam) also depend on the choice of coordinate system.

In order to derive the invariant model of beam, we modify the definition of the bending moments. Using axial stress,

$$N = \langle \sigma_{11}^{\varepsilon} \rangle , \tag{A18}$$

we introduce the following quantities

$$M_{\alpha}^{0} = M_{\alpha} - \tau \langle y_{\alpha} \rangle N \text{, where } \tau = \frac{mesY}{m}, \tag{A19}$$

which will be called normalized moments to distinguish them from the normally used moments M_{α}. The number τ in (A19) is taken in such way, that the equality $\tau\langle y_{\alpha} + h_{\alpha} \rangle = \tau \langle y_{\alpha} \rangle + h_{\alpha}$ be satisfied.

The normalized moments are invariants (do not depend on the choice of the coordinate system). Really, in the "new" coordinate system we have

$$M_{\alpha}^{0}(h) = M_{\alpha}(h) - \tau \langle y_{\alpha} + h_{\alpha} \rangle N_{\alpha\beta}$$

$$= M_{\alpha} + h_{\alpha} N_{\alpha} - \tau \langle y_{\alpha} \rangle N - \tau \langle h_{\alpha} \rangle N$$

$$= M_{\beta} - \tau \langle y_{\alpha} \rangle N + h_{\alpha} (1 - \tau \frac{m}{mesY})$$

$$= M_{\alpha} - \tau \langle y_{\alpha} \rangle N = M_{\alpha}^{0} .$$

The homogenized constitutive equations (3.2.45), (3.2.46) of a beam can be written in the form (we omit here the terms corresponding to torsion),

$$N(h) = A(h)e + A_{\alpha}^{1}(h)\rho_{\alpha},$$

$$M_{\alpha}^{0}(h) = A_{\alpha}^{1}(h)e + A_{\alpha\beta}^{2}(h)\rho_{\beta} - \tau\langle y_{\alpha} \rangle (A(h)e + A_{\alpha}^{1}(h)\rho_{\alpha}) .$$

These equations can be written as

$$N(h) = A(e + h_{\alpha}\rho_{\alpha}) + A_{\alpha}^{1}(h)\rho_{\alpha}, \tag{A20}$$

$$M_{\alpha}^{0}(h) = A_{\alpha}^{1}(e + h_{\alpha}\rho_{\alpha}) + A_{\alpha\beta}^{2}\rho_{\beta} - \tau\langle y_{3} \rangle [A(e + h_{\alpha}\rho_{\alpha}) + A_{\alpha\beta}^{1}\rho_{\beta}] .$$

Equations (A20), considered with respect to the variables

$$N\,,\;M^0_\alpha,\;e_\alpha + h_\alpha \rho_\alpha,\;\rho_\alpha, \tag{A21}$$

are invariant with respect to the change (A12) of coordinate system.

Homogenized equations of equilibrium (3.2.49) can be written in the form,

$$N_{,1x} = \langle f_1 \rangle + \langle g_1 \rangle_S\,, \tag{A22}$$

$$M^0_{\alpha,1x1x} + (\langle f_1 y_\beta \rangle + \langle g_1 y_\beta \rangle)_{,1x}$$

$$= \langle f_\alpha \rangle + \langle g_\alpha \rangle_S - \tau \langle y_\alpha \rangle (\langle f_1 \rangle + \langle g_1 \rangle_S),$$

invariant with respect to change (A12) of the coordinate system.

Thus, we write all the equations forming the homogenized model in a form invariant with respect to change (A12) of coordinate system.

The simplest form of homogenized constitutive equations of a beam

Let us analyze the formulas (A15), (A16). It is seen, that the bending stiffness $A^2_{\alpha\alpha}(h)$ takes the minimal value when,

$$h_\alpha = -\frac{A^1_\alpha}{A} \quad (\alpha = 1, 2)\,. \tag{A23}$$

It is seen from (A16) that h_1 and h_2 determined by (A23) give, simultaneously, zero value to the coupling stiffness $A^1_\alpha(h)$. The equations (A23) formally introduce the neutral axis $\{y_1 = h_1,\; y_2 = h_2\}$ for composite beam. The neutral axis can correspond to no real object. The similar situation one can meet considering a cylindrical beam with cylindrical holes.

Torsional stiffness of a beam

Consider cellular problem (3.2.42) used to compute the torsional stiffness of a beam. In the "new" coordinate system, it takes the form.

$$\begin{cases} [c_{ijmn}(\mathbf{y})X^3_{m,ny} + y_\Gamma s_\gamma c_{ij\gamma 1}(\mathbf{y})]_{,j} = 0 \text{ in } Y, \\ [c_{ijmn}(\mathbf{y})X^3_{m,ny} + y_\Gamma s_\gamma c_{ij\gamma 1}(\mathbf{y})]n_j = 0 \text{ on } S, \\ \mathbf{X}^3(\mathbf{y}) \text{ is periodic in } y_1, y_2 \text{ with periodicity cell } T. \end{cases} \tag{A24}$$

Solution $\mathbf{X}^3(h, \mathbf{y})$ of problem (A24), with regard for Proposition 3.4, is related with solution $\mathbf{X}^3(\mathbf{y})$ in the "old" coordinate system as follows:

$$\mathbf{X}^3(h, \mathbf{y}) = \mathbf{X}^3(\mathbf{y}) - h_\Gamma s_\gamma y_\gamma \mathbf{e}_1 . \tag{A25}$$

Computer the term $c_{\mathrm{B}1mn}(\mathbf{y})X^3_{m,ny} + (y_\mathrm{B} + h_\mathrm{B})s_\gamma c_{\gamma 1\gamma 1}$, which is multiplier in formula (3.2.48) for computing of torsional stiffness B. Substituting $\mathbf{X}^3(h, \mathbf{y})$ in accordance with (A25), we find that this term is equal to

$$c_{\mathrm{B}1mn}(\mathbf{y})X^3_{m,ny} + c_{\mathrm{B}1mn}(\mathbf{y})(-y_\Gamma s_\gamma h_\Gamma \mathbf{e}_1)_{m,ny} \tag{A26}$$

$$+y_\Gamma s_\gamma c_{\mathrm{B}1\gamma 1} + h_\Gamma s_\gamma c_{\mathrm{B}1\gamma 1}$$

$$= c_{\mathrm{B}1mn}(\mathbf{y})X^3_{m,ny} + y_\Gamma s_\gamma c_{\mathrm{B}1\gamma 1} + h_\Gamma s_\gamma c_{\mathrm{B}1\gamma 1} - c_{\mathrm{B}1\gamma 1}(\mathbf{y})y_\Gamma s_\gamma h_\Gamma$$

$$= c_{\mathrm{B}1mn}(\mathbf{y})X^3_{m,ny} + y_\Gamma s_\gamma c_{\gamma 1\gamma 1} .$$

Here we use the equality $c_{\gamma 11\gamma} = c_{\gamma 1\gamma 1}$.

Then, in the "new" coordinate system

$$B(h) = \langle (y_\beta + h_\beta)s_\beta [c_{\mathrm{B}1mn}(\mathbf{y})X^3_{m,ny} + y_\Gamma s_\gamma c_{\mathrm{B}1\gamma 1}] \rangle \tag{A27}$$

$$= B + h_\beta s_\beta \langle c_{\mathrm{B}1mn}(\mathbf{y})X^3_{m,ny} + y_\Gamma s_\gamma c_{\mathrm{B}1\gamma 1} \rangle .$$

The last term in (A27) is equal to zero. It can be proved by multiplying of the differential equation from (A24) by 1 and integrating by parts. As a result,

$$B(h) = B\ .$$

It means that the torsional stiffness is an invariant.

The torsional moment is determined by the formula

$$M = \langle \sigma_{31}^{\varepsilon} y_2 - \sigma_{21}^{\varepsilon} y_3 \rangle\ .$$

In the "new" coordinate system

$$M(h) = \langle \sigma_{31}^{\varepsilon}(y_2 + h_2) - \sigma_{21}^{\varepsilon}(y_3 + h_3) \rangle$$

$$= M + \langle \sigma_{31}^{\varepsilon} \rangle h_2 - \langle \sigma_{21}^{\varepsilon} \rangle h_3\ .$$

In accordance with Proposition 3.6, $\langle \sigma_{31}^{\varepsilon} \rangle = 0$ and $\langle \sigma_{32}^{\varepsilon} \rangle = 0$. Then

$$M(h) = M\ .$$

Thus, the torsion moment is an invariant.

Appendix B. Critical Loads for Honeycomb Filled and Wafer Plates

In this section, we give examples addressed to one interesting only in the application of the theory developed in this book. The author selects for this section two examples, which give imagine about the possibilities of the theory developed with minimal mathematical computations.

Honeycomb filled plate

Consider a right-angle honeycomb filled plate (see Fig. B1). Let edges $x_1 = 0, L$ be free, edges $x_2 = 0, L$ be simply supported. At edges $x_2 = 0, L$, the plate is subjected to in-plane loads P parallel to Ox_2-axis; Fig.B1.

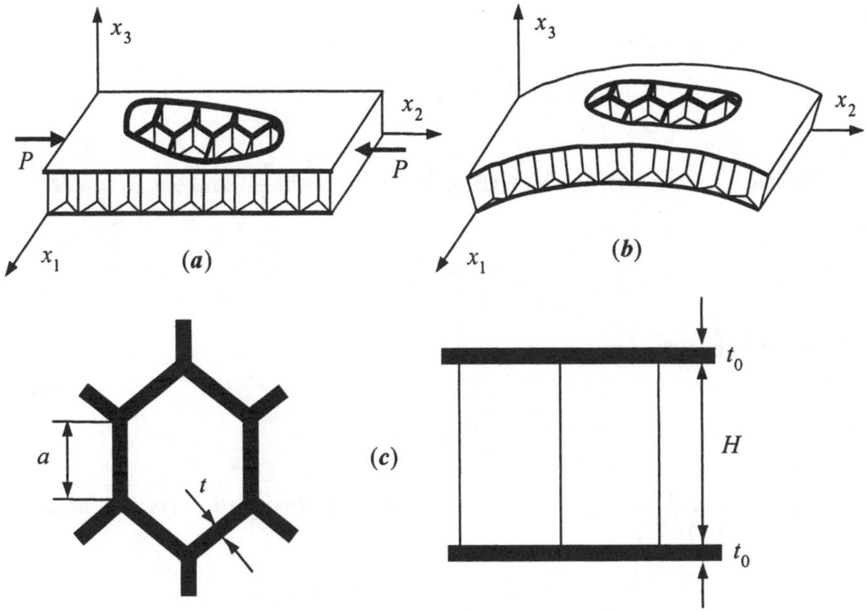

Fig. B1. A honeycomb filled plate: (***a***) before buckling; (***b***) after buckling; and (***c***) geometrical parameters of the periodicity cell of plate

In the case under consideration, the initial homogenized stresses are determined independently on the in-plane stiffnesses of plate. They have the form: $\sigma_{22} = \frac{P}{L}$, and $\sigma_{ij} = 0$ if $ij \neq 22$.

We consider the problem of loss of stability for the plate under consideration and find the critical load (the load inducing a loss of stability of plate). A direct analysis of plate shown in Fig. B1 is impossible even with modern computers, because the total number of constitutive elements of plate is very large. The homogenization approach allows analyze the problem easily.

For the plate under consideration, coupling stiffness $A^1_{2222} = 0$ by virtue of symmetry of the plate. By virtue of $A^1_{2222} = 0$, equations (2.2.66) transform into one equation with respect the normal deflection w,

$$A^2_{2222} w^{\mathrm{IV}} = \frac{P}{L} \lambda w'' , \tag{B1}$$

with boundary conditions

$$w(0) = w''(0) = w(1) = w''(1) = 0 . \tag{B2}$$

The eigenfunction for problem (B1), (B2) is $w = \sin \frac{\pi x}{L}$, and eigenvalue is

$$\frac{P}{L} = \left(\frac{\pi}{L} \right)^2 A^2_{2222} .$$

Thus, the critical load is

$$P = \frac{\pi^2}{L} A^2_{2222} . \tag{B4}$$

The homogenized bending stiffness of honeycomb filled plate is (see Kalamkarov and Kolpakov, 1997)

$$A^2_{2222} = \frac{E_0}{1 - \nu_0^2} \left(\frac{H^2 t_0^2}{2} + H t_0^2 + \frac{2}{3} t_0^3 \right) + \frac{\sqrt{3}}{48} \frac{E H^3 t}{a} ,$$

where E_0 and ν_0 are Young's modulus and Poisson's ratio of carrier layers, E is Young's modulus of filler; the geometrical parameters H, t_0, t, and a are shown in Fig.B1(*c*).

Wafer plate

Consider a right-angle wafer plate (Fig. B2). Edges $x_1 = 0, L$ are free, and edges $x_2 = 0, L$ are simply supported. At edges $x_2 = 0, L$, in-plane load P is applied to the plate. The initial homogenized in-plane stresses are: $\sigma_{22} = \frac{P}{L}$, and $\sigma_{ij} = 0$ if $ij \neq 22$.

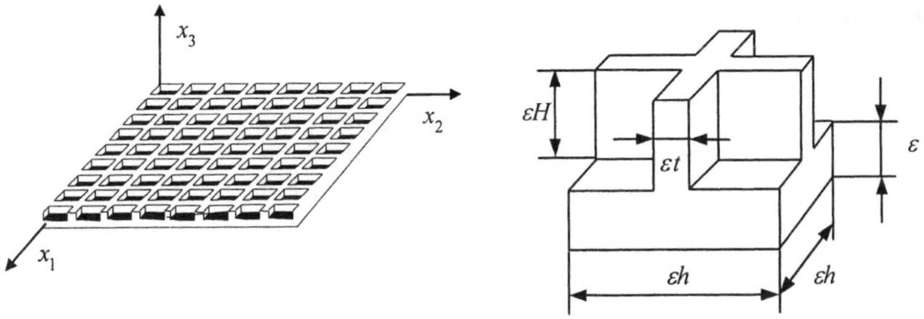

Fig. B2. A wafer plate and geometrical parameters of periodicity cell

The homogenized stiffnesses of wafer plate are the following (see Kalamkarov and Kolpakov, 1997):

(B5)

$$A^0_{2222} = E\left(\frac{1}{1-\nu^2} + F\right),$$

$$A^1_{2222} = ES\,,$$

$$A^2_{2222} = E\left(\frac{1}{12(1-\nu^2)} + J\right),$$

where

$$F = \frac{Ht_1}{h_1}, \quad S = \frac{(H^2 + H)t_1}{2h_1},$$

$$J = \frac{(4H^3 + 6H^2 + 3H)t_1}{12h_1},$$

E and ν are Young's modulus and Poisson's ratio of the material from which the plate is made; the geometrical parameters H, h, and t are shown in Fig. B2.

Stiffnesses (B5) were computed (Kalamkarov and Kolpakov, 1997) in the coordinate system in which $A^1_{2222} \neq 0$. Equation (B1) and formula (B3) are valid in the coordinate system in which $A^1_{2222} = 0$. We can pass to a new coordinate system in which $A^1_{2222} = 0$ (see Appendix A). In the new coordinate system, the bending stiffness is

$$A^2_{2222} = E\left(\frac{1}{12(1-\nu^2)} + J - \frac{S^2}{\dfrac{1}{1-\nu^2} + F} \right). \tag{B5}$$

Computing of local stresses

Local stresses σ^{loc}_{ij} in plate are the sum of the initial local stresses [see (2.8.8)],

$$\sigma^*_{ij} = S_{ijmn}(\mathbf{x}/\varepsilon)V_{\alpha,\beta x}(x_1, x_2) + R_{ijmn}(\mathbf{x}/\varepsilon)u^{(0)}_{3,\alpha x\beta x}(x_1, x_2),$$

and additional local stresses [see (2.2.44)],

$$\sigma^\varepsilon_{ij} = S_{ijmn}(\mathbf{x}/\varepsilon)V_{\alpha,\beta x}(x_1, x_2) + R_{ijmn}(\mathbf{x}/\varepsilon)u^{(0)}_{3,\alpha x\beta x}(x_1, x_2).$$

Formulas for tensors of local stress concentrations S_{ijmn} (corresponding to in-plane deformations) and R_{ijmn} (corresponding to bending) for honeycomb, wafer, and some other kinds of plates can be found in the book by Kalamkarov and Kolpakov (1997).

References

Almgren, (1985). An isotropic three-dimensional structure with Poisson's ratio=−1. *J. Elasticity*, 15, 427–430.

Annin, B.D., Kalamkarov, A.L., Kolpakov, A.G. and Parton, V.Z. (1993). *Computation and Design of Composite Materials and Structural Elements.* Nauka, Novosibirsk (in Russian).

ANSYS Modeling and Meshing Guide, (1992). 3rd Edition. SAS IP Inc.

Artola, M. and Duvaut, G. (1977). Homogeneisation d'une plaque renforcee. *Comp Rend. Acad. Sci. Paris.*, Ser. A. 288, 775–778.

Babushka, I (1976). Solution of interface problem by homogenization. Parts I,II. *SIAM J. Math. Anal.*, 7, 603–645.

Bakhvalov, N.S. (1978). Averaged characteristics of bodies with periodic structures. *Doklady Acad. Nauk SSSR*, 218(5), 1046–1048.

Bakhvalov, N.S. and Panasenko, G.P. (1989). *Homogenization: Averaging Processes in Periodic Media.* Kluwer, Dordrecht (Russian edition, 1982, Nauka, Moscow).

Banhart, J., Ashby, M. and Fleck, N., Eds. (2001). *Cellular Metals and Metal Foaming Technology.* MIT Publ., Bremen.

Bathe, K.J. (1982). *Finite Elements Procedures in Engineering Analysis.* Prentice-Hall, Englewood Cliffs, NJ.

Bendsøe, M.P. (1995). *Optimization of Structural Topology, Shape and Material.* Springer, Heidelberg.

Bensoussan, A., Lions, J.L. and Papanicolaou, G. (1978). *Asymptotic Analysis for Periodic Structures.* North-Holland, Amsterdam.

Beran M. and Molyneux, J. (1966). Use of classical variational principles to determine bounds for the effective bulk modulus in heterogeneous media. *Q. Appl. Math.*, 24.

Berdichevskii,V.L. (1975). Spatial averaging of periodic structures. *Doklady Acad. Nauk SSSR*, 222(3), 565–567.

Boccardo, L. and Marcellini, P. (1976). Sulla convergenza della soluzioni di disequazioni variazionali. *Ann. Mat. Pura Appl.*, 110. 137–159.

Broutman, L.J. and Krock, R.H., Eds. (1974). *Composite Materials.* In eight volumes. Academic Press, NY, London (Russian edition, 1978, Mir, Moscow).

Bryan G.H. (1889). On the stability of elastic systems. *Proc. Cambridge Philos. Soc.*, 6, 199–210.

Caillerie, D. (1984). Thin elastic and periodic plates. *Math. Methods Appl. Sci.*, 6, 159–191.

Chou, T.-W. and Ko, F.K., Eds. (1989). *Textile Structural Composites.* Elsevier, Amsterdam (Russian edition. 1991, Mir, Moscow).

Christensen, R.M. (1979). *Mechanics of Composite Materials.* John Wiley&Sons, New York (Russian edition, 1982, Mir, Moscow).

Ciarlet, P.G. (1990). *Plates and Junctions in Elastic Multi-Structures. An Asymptotic Analysis.* Macon, Paris.

Ciarlet, P.G. and Destuyner, P. (1979). A justification of the two-dimensional linear plate model. *J. Mecanique*, 18, 315–344.

Cioranescu, D. and Donato, P. (2000). *An Introduction to Homogenization.* Oxford University Press, Oxford.

Courant, R.S. and Hilbert D. (1953). *Methods of Mathematical Physics.*.Vol.1. John Wiley&Sons, New York.

De Giorgi, E. and Spagnolo, S. (1973). Sulla converzenca delli integrali dell energia per operatori ellittico del secondo ordine. *Boll. Unione Mat. Ital.*, 8, 391–411.

Duvaut, G. (1976). Analyse fonctionelle et mechanique des mulieux continus. *Proc. 14th IUTAM Congress*, Delft. Amsterdam. 119–132.

Duvaut, G. (1977). Comportement microscopique d'une plate perfore periodiqumente. In: *Singular Perturbations and Boundary Layers.* Lecture Notes in Mathematics. V.594, Springer, Berlin.

Ekeland, I. and Temam, R. (1976). *Convex Analysis and Variational Problems.* North-Holland, Amsterdam (Russian edition, 1979, Mir, Moscow).

Euler, L. (1759). Sur la force des colonnes. *Histoire de l'Academie Royal des sciences et belle-lettres avec les Memoires, tires des Registres de cette Academie.* 1759, T.13. 252–282.

Fichera, G. (1972). *Existence Theorems in Elasticity.* Springer–Verlag, Berlin, New York.

Francfort, G.A. and Murat, F. (1986). Homogenization and optimal bounds in linear elasticity. *Arch. Rat. Mech. Anal.*, 94. 307–334.

Gajewski, H., Gröger, K. and Zacharias, K. (1974). *Nichtlineare Operatorleichngen und Operatordifferential–gleichungen.* Academie–Verlag, Berlin (Russian edition, 1978, Mir, Moscow).

Gibson, L.J. and Ashby, M.F. (1997). *Cellular Solids. Structures and Properties.* Cambridge University Press, Cambridge.

Grigoluk, E.I and Fil'shtinskii, L.A. (1970). *Perforated Plates and Shells.* Nauka, Moscow (in Russian).

Hashin Z. and Shtrikman S. (1963). A variational approach to the theory of the elastic behavior of multiphase materials. *J. Mech. Phys. Solids*, 11(2).

Haug, E.J., Choi, K.K. and Komkov, V. (1986). *Design Sensitivity Analysis of Structural Systems.* Academic Press, Orlando, San Diego, New York (Russian edition, 1988, Mir, Moscow).

Hill, R. (1952). The elastic behavior of a crystalline aggregate of anisotropic cubic crystals. *Proc. Phys. Soc.*, A65, 389.

Hill, R. (1958). A general theory of uniqueness and stability in elastic-plastic solids. *J. Mech. Phys. Solids*, 6(3), 236–249.

Hrennikoff, A. (1941). Solution of problem of elasticity by framework method. *J. Appl. Mech.*, A169–A175.

Introduction to MSC/NASTRAN. (1994). MacNeal–Schewdler Corp.

Jikov, V.V, Kozlov, S.M. and Oleinik O.A. (1994). *Homogenization of Differential Operators and Integral Functionals.* Springer–Verlag, Heidelberg.

Jones, R.M. (1975). *Mechanics of Composite Materials.* Scripta Book Co., Washington, DC.

Kalamkarov, A.L. and Kolpakov, A.G. (1997). *Analysis, Design and Optimization of Composite Structures.* John Wiley&Sons, Chichester, New York.

Kelly, A. and Rabotnov, Yu.N., Eds. (1988). *Handbook of Composites.* North-Holland, Amsterdam.

Kohn, R.V. and Vogelius, M. (1984). A new model for thin plates with rapidly varying thickness. *Int. J. Solids Struct.*, 20, 333–350.

Kolmogorov, A.N. and Fomin, S.V. (1970). *Introductory Real Analysis*, Prentice-Hall, Englewood Cliffs, NJ.

Kolpakov, A.G. (1981). Stability of inhomogeneous plate with unilateral restrictions. *Dinamika sploshnoi sredy.* Lavrent'ev Inst. hydrodinamiki SO AN SSSR. Novosibirsk. 53, 31–43.

Kolpakov, A.G. (1985). Determining of the averaged characteristics of elastic frameworks. *J. Appl. Math. Mech.*, 49, 969–977.

Kolpakov, A.G. (1987). Averaging in problems of the bending and oscillation of stressed inhomogeneous plate. *J. Appl. Math. Mech.*, 51, 44–49.

Kolpakov, A.G. (1989). Rigidity characteristics of stressed inhomogeneous media. *Mech. Solids.* (English transl. of *Izvestiya Akad. Nauk SSSR. MTT*), N 3. 66–73.

Kolpakov, A.G. (1990). On dependence of velocity of elastic waves in composite media on initial stresses. *Second World Cong. Comp. Mech., Stuttgart, FRG. Extended Abstracts of Lectures*, 453–456.

Kolpakov, A.G. (1991). Calculation of the characteristics of thin elastic rods of periodic structure. *J. Appl. Math. Mech.*, 57, 440–448.

Kolpakov, A.G. (1992a). On problem of beam with initial stress. *Appl. Mech. Tech. Phys.* (English transl. of *Prikl. Mekh. Teh. Fiz. USSR*), N 6, 139–144.

Kolpakov, A.G. (1992b). On the dependence of velocity of elastic waves in composite media on initial stresses. *Comp. Struct.* 44, 97–101.

Kolpakov, A.G. (1994). Stiffnesses of elastic cylindrical beams. *J. Appl. Math. Mech.*, 58, 102–109.

Kolpakov, A.G. (1995a). On the problem of plate with initial stresses. *Mech. Solids* (English transl. of *Izvestiya Akad. Nauk SSSR. MTT*), N 3. 179–187.

Kolpakov, A.G. (1995b). Asymptotic problem stability of beams. Buckling under torsion. *Appl. Mech. Tech. Phys.* (English transl. of *Prikl. Mekh. Teh. Fiz. USSR*), N 6. 133–141.

Kolpakov, A.G. (1998a). Application of homogenization method to justification of 1-D model for beam of periodic structure having initial stresses. *Int. J. Solids Struct.*, 35(22), 2847–2859.

Kolpakov, A.G. (1998b). Variational principles for stiffnesses of a non-homogeneous beam. *J. Mech. Phys. Solids*, 46(6), 1039–1053.

Kolpakov, A.G. (1999). Variational principles for stiffnesses of a non-homogeneous plate. *J. Mech. Phys. Solids*, 47, 2075–2092

Kolpakov, A.G. (2000). Homogenized model for plate periodic structure with initial stresses. *Int. J. Eng. Sci.*, 38(18), 2079–2094

Kolpakov, A.G. (2001). On the calculation of rigidity characteristics of the stressed constructions. *Int. J. Solids Struct.*, 38(15), 2469–2485

Kolpakov, A.G. and Sheremet, I.G (1997). Asymptotic model of membrane and string. *J. Appl. Math. Mech.*, 61, 845–853.

Kozlov, S.M. (1978). Averaging of random structures. *Doklady Acad. Nauk SSSR*, 241(5), 1016–1019.

Kozlova, M.V. (1989). Averaging of the three-dimensional elasticity problem for a thin inhomogeneous rod. *Vestnik Moscowskogo Universiteta, Ser.1, Math. Mech.*, 5. 6–10.

Ladizenskaya, O.A. (1973). *Boundary Value Problems of Mathematical Physics.* Nauka, Moscow (in Russian).

Lakes, R. (1987). Foam structures with negative Poisson's ratio. *Science*, 235 (Feb.), 1038.

Langhaar, H.L. (1962). *Energy Method in Engineering.* John Wiley&Sons, New York.

Lions, J–L. (1978). Notes on some computational aspects of homogenization method for composite materials. In: *Computational Methods in Mathematics, Geophysics and Optimal Control.* Nauka, Novosibirsk.

Lions, J–L. and Magenes, E. (1972). *Non-Homogeneous Boundary Value Problems and Applications.* Springer–Verlag, Berlin, New York (Russian edition. 1971, Mir, Moscow).

Love, A.E.H. (1929). *A Treatise on the Mathematical Theory of Elasticity.* Cambridge University Press, London (Russian edition, 1935, GITTL, Moscow).

Marcellini, P. (1973). Su una convergenza di funzioni convesse. *Bull. Unione Mat. Ital.*, 8(1).

Marcellini, P. (1975). Un teorema di passagio de limite per la somma di convesse. *Bull. Unione Mat. Ital.*, 4, 107–124.

Marguerre, K. (1938). Die Behandlung der Stabilitätsproblemen mit Hilfe der energetishen Methoden. *ZAMM*, 18(1), 57–73.

Mignot, F., Puel, J–P. and Suquet, P–M. (1980). Homogenization and bifurcation of perforated plates. *Int. J . Eng. Sci.*, 18. 409–414.

Milton, G.V. and Kohn, R.V. (1988). Variational bounds on the effective moduli of anisotropic composites. *J. Mech.Phys. Solids*, 36. 597–629.

Nemat–Nasser, S. and Hori, M. (1993). *Micromechanics: Overall Properties of Heterogeneous Materials.* North-Holland, Amsterdam.

Noor, A.K. (1988). Continuum modeling for repetitive structures. *Appl. Mech. Rev.*, 41(7), 285–296.

Novozilov, V.V. (1948). *Foundation of Non-Linear Elasticity Theory.* Gostehizdat, Moscow-Leningrad (in Russian).

Oleinik, O.A., Iosifian, G.A. and Shamaev, A.S. (1990). *Mathematical Problems of Theory of Nonhomogeneous Media.* Moscow State University Publ., Moscow (in Russian).

Panasenko, G.P. (1978). High-order asymptotic for solution of equations with fast oscillating coefficients. *Doklady Acad. Nauk SSSR*, 240(6), 1293–1296.

Panasenko, G.P. and Reztsov, M.V. (1987). Homogenization of the three–dimensional elasticity problem for an inhomogeneous plate. *Doklady Akad. Nauk SSSR*, 294, 1061–1065.

Pipes, L.A. (1963). *Matrix Method in Engineering*, Prentice-Hall, Englewood Cliffs, NJ.

Prager, W. (1947). The general variational principle of the theory of structural stability. *Q. Appl. Math*, 4(4), 378–384.

Pshenichnov, G.I. (1982). *Theory of Thin Elastic Network Plates and Shells.* Nauka, Moscow (inRussian).

Reuss, A. (1929). Berechnung der fliebgrenze von mischkristallen auf grund der platizitatsbedingung fur einkristalle. *ZAMM*, 9(1), 49–58.

Samarskii, A. A. (1977). *Theory of Difference Schemes.* Nauka, Moscow (in Russian).

Sanchez–Palencia, E. (1970). Equations aux deivees partielles. Solutions periodiques par repport aux variables d'espace et application. *Comp. Rend. Acad. Sci. Paris*, ser. A, 1129–1132.

Sanchez–Palencia, E. (1980). *Non-Homogeneous Media and Vibration Theory.* Lect. Notes in Physics, 127. Springer, Berlin (Russian edition, 1994, Mir, Moscow).

Sendeckyj, G.P. (1974). Elastic behaviour of composite materials. In: *Composite Materials. Vol.2. Mechanics of Composite Materials.* Academic Press, Ney York.

Spagnolo, S. (1968). Sul limite delle soluzioni di equazioni paraboliche et elittiche. *Ann. Scuola Norm. Sup. Pisa*, 22. 577–597.

Timoshenko, S. (1953). *History of Strength of Materials.* McGraw–Hill, New York.

Timoshenko, S. (1961). *Theory of Elastic Stability.* 2nd edition, McGraw–Hill, New York.

Timoshenko, S. and Goodier, J.N. (1970). *Theory of Elasticity.* McGraw–Hill, New York (Russian edition, 1975, Nauka, Moscow).

Timoshenko, S. and Woinowsky–Krieger, K. (1959). *Theory of Plates and Shells.* McGraw-Hill, New York (Russian edition, 1963, Fizmatgiz, Moscow).

Trabucho, L., and Viaño J.M. (1987). Derivation of generalized models for linear elastic beams by asymptotic expansion method. In: Ciarlet P.G. and Sanchez-Palencia E., Eds. *Application of Multiple Scaling in Mechanics.* Masson, Paris, 129–148.

Trabucho, L., and Viaño J.M. (1996). *Mathematical Modeling of Rods. Handbook of Numerical Analysis*, Elsevier (P.G. Ciarlet and J.L. Lions, eds.).

Trefftz, E. (1930).Über die Ableitung der Stabilitäs-Kriterien des elastichen Gleichgewichtes aus der Elastitätstheorie endlicher Deformationen. *Proc. 3rd Int. Congr. Appl. Mech.*, 44–50.

Trefftz, E. (1933). Zur theorie der stabilität des elastichen gleichgewichtz. *ZAMM*, 13(2), 160–165.

Tutek, Z. and Aganovich, I. (1987). A justification of the one-dimensional linear model of elastic beam. *Math. Methods Appl. Sci.* 8, 502–515.

Vinson, J.R. (1989). *The Behavior of Thin-Walled Structures: Beams, Plates and Shells.* Kluwer, Dordrecht.

Voight, W. (1928). *Lehrbuch der Kristalphysik.* Tubner, Berlin.

Washizu, K. (1982). *Variational Methods in the Theory of Elasticity and Plasticity.* Pergamon Press, Oxford (Russian edition, 1987, Mir, Moscow).

Willis, J.R. (1981). Variational and related methods for the overall properties of anisotropic composites. *Appl. Mech. Rev.*, 21, 1–78.

www.abaqus.com (electronic)

Yeh, R.H.T. (1970). Variational principles of elastic moduli of composite materials. *J. Appl. Phys.*, 41(8).

Zikov, V., Kozlov, S., Oleinik O., and Ngoan, K. (1979). Averaging and G-convergence of differential operators. *Russ. Math. Surv.*, 34, 69–147.

Index

Foundations of Engineering Mechanics

Series Editors: Vladimir I. Babitsky, Loughborough University
Jens Wittenburg, Karlsruhe University

Further volumes of this series can be found on our homepage: springeronline.com

Landa, P. S.
Regular and Chaotic Oscillations, 2001
ISBN 3-540-41001-5

Muravskii, G.
Mechanics of Non-Homogeneous and Anisotropic Foundations 2001
ISBN 3-540-41631-5

Babitsky, V. I., Krupenin, V. L.
Vibration of Strongly Nonlinear Discontinuous Systems, 2001
ISBN 3-540-41447-9

Gorshkov, A. G., Tarlakovsky, D. V.,
Transient Aerohydroelasticity of Spherical Bodies, 2001
ISBN 3-540-42151-3

Lurie, A. I.
Analytical Mechanics, 2002
ISBN 3-540-42982-4

Manevitch, L. I., Andrianov, I. V., Oshmyan, V. G.
Mechanics of Periodically Heterogeneous Structures, 2002
ISBN 3-540-41630-7

Slepyan, L. I.
Models and Phenomena in Fracture Mechanics, 2002
ISBN 3-540-43767-3

Nagaev, R. F.
Dynamics of Synchronising Systems, 2003
ISBN 3-540-44195-6

Svetlitsky, V. A.
Statistical Dynamics and Reliability Theory for Mechanical Structures, 2003
ISBN 3-540-44297-9

Neimark, J. I.
Mathematical Models in Natural Science and Engineering, 2003
ISBN 3-540-43680-4

Babitsky, V. Shipilov, A.
Resonant Robotic Systems, 2003
ISBN 3-540-00334-7

Le xuan Anh
Dynamics of Mechanical Systems with Coulomb Friction, 2003
ISBN 3-540-00654-0

Perelmuter, A., Slivker, V. I.
Numerical Structural Analysis, 2003
ISBN 3-540-00628-1

Andrianov, I. V., Awrejcewicz, J., Manevitch, L. I. (Eds.)
Asymptotical Mechanics of Thin-Walled Structures, 2004
ISBN 3-540-40876-2

Shorr, B. F.
The Wave Finite Element Method, 2004
ISBN 3-540-41638-2

Ginevsky, A., Vlasov, Y. V., Karasov, R. K.
Acoustic Control of Turbulent Jets, 2004
ISBN 3-540-20143-2

Svetlitsky, V. A.
Engineering Vibration Analysis 1, 2004
ISBN 3-540-20658-2

Svetlitsky, V. A.
Engineering Vibration Analysis 2, 2004
ISBN 3-540-20782-1

Kolpakov, A. G.
Stressed Composite Structures, 2004
ISBN 3-540-40790-1

Printing: Strauss GmbH, Mörlenbach
Binding: Schäffer, Grünstadt